# Preface

The present book is about differential geometry of space curves and surfaces. The formulation and presentation are largely based on a tensor calculus approach, which is the dominant trend in the modern mathematical literature of this subject, rather than the geometric approach which is usually found in some old style books. The book is prepared, to some extent, as part of tutorials about topics and applications related to tensor calculus. It can therefore be used as part of a course on tensor calculus as well as a textbook or a reference for an intermediate-level course on differential geometry of curves and surfaces.

Apart from general background knowledge in a number of mathematical branches such as calculus, geometry and algebra, an important requirement for the reader and user of this book is familiarity with the terminology, notation and concepts of tensor calculus at reasonable level since many of the notations and concepts of differential geometry in its modern style are based on tensor calculus.

The book contains a mathematical background section in the first chapter to outline some important pre-required mathematical issues. However, this section is restricted to materials related directly to the contents of differential geometry of the book and hence the reader and user should not expect this mathematical background section to be comprehensive in any way. General mathematical knowledge, plus possible consultation of mathematical textbooks related to other disciplines of mathematics when needed, should therefore be considered.

The book is furnished with an index in the end of the book as well as sets of exercises in the end of each chapter to provide useful revisions and practice. To facilitate linking related concepts and parts, and hence ensure better understanding of the provided materials, cross referencing is used extensively throughout the book where these referrals are hyperlinked in the electronic version of the book for the convenience of the ebook users. The book also contains a considerable number of graphic illustrations to help the readers and users to visualize the ideas and understand the abstract concepts.

The materials of differential geometry are strongly interlinked and hence any text about the subject, like the present one, will face the problem of arranging the materials in a natural order to ensure gradual development of concepts. In this book we largely followed such a scheme. However, this is not always possible and hence in some cases references are provided to materials in later parts of the book for concepts needed in earlier parts. Nevertheless, in most cases brief definitions of the main concepts are provided in the first chapter in anticipation of more detailed definitions and investigations in the subsequent chapters.

Regarding the preparation of the book, everything is made by the author including all the graphic illustrations, indexing, typesetting, book cover, as well as overall design. In this regard, I should acknowledge the use of LaTeX typesetting package and the LaTeX based document preparation package LyX which facilitated the typesetting and design of the book substantially.

Taha Sochi, London, March 2017

# Contents

| | |
|---|---|
| Table of Contents | 2 |
| Nomenclature | 5 |
| **1 Preliminaries** | **8** |
| 1.1 Differential Geometry | 8 |
| 1.2 General Remarks, Conventions and Notations | 8 |
| 1.3 Classifying the Properties of Curves and Surfaces | 10 |
|     1.3.1 Local versus Global Properties | 10 |
|     1.3.2 Intrinsic versus Extrinsic Properties | 11 |
| 1.4 General Mathematical Background | 12 |
|     1.4.1 Geometry and Topology | 12 |
|     1.4.2 Functions | 20 |
|     1.4.3 Coordinates, Transformations and Mappings | 23 |
|     1.4.4 Intrinsic Distance | 29 |
|     1.4.5 Basis Vectors | 29 |
|     1.4.6 Flat and Curved Spaces | 30 |
|     1.4.7 Homogeneous Coordinate Systems | 32 |
|     1.4.8 Geodesic Coordinates | 32 |
|     1.4.9 Christoffel Symbols for Curves and Surfaces | 34 |
|     1.4.10 Riemann-Christoffel Curvature Tensor | 36 |
|     1.4.11 Ricci Curvature Tensor and Scalar | 38 |
| 1.5 Exercises | 39 |
| **2 Curves in Space** | **43** |
| 2.1 General Background about Curves | 43 |
| 2.2 Mathematical Description of Curves | 46 |
| 2.3 Curvature and Torsion of Space Curves | 51 |
|     2.3.1 Curvature | 52 |
|     2.3.2 Torsion | 54 |
| 2.4 Geodesic Torsion | 55 |
| 2.5 Relationship between Curve Basis Vectors and their Derivatives | 56 |
| 2.6 Osculating Circle and Sphere | 57 |
| 2.7 Parallelism and Parallel Propagation | 59 |
| 2.8 Exercises | 60 |
| **3 Surfaces in Space** | **65** |
| 3.1 General Background about Surfaces | 65 |
| 3.2 Mathematical Description of Surfaces | 71 |
| 3.3 Surface Metric Tensor | 76 |

|  |  | 3.3.1 Arc Length . . . . . . . . . . . . . . . . . . . . . . . . . . . | 79 |

       3.3.1  Arc Length . . . . . . . . . . . . . . . . . . . . . . . . . . . 79
       3.3.2  Surface Area . . . . . . . . . . . . . . . . . . . . . . . . . . 80
       3.3.3  Angle Between Two Surface Curves . . . . . . . . . . . . . 82
  3.4  Surface Curvature Tensor . . . . . . . . . . . . . . . . . . . . . . . . 83
  3.5  First Fundamental Form . . . . . . . . . . . . . . . . . . . . . . . . . 86
  3.6  Second Fundamental Form . . . . . . . . . . . . . . . . . . . . . . . 89
       3.6.1  Dupin Indicatrix . . . . . . . . . . . . . . . . . . . . . . . . 93
  3.7  Third Fundamental Form . . . . . . . . . . . . . . . . . . . . . . . . 94
  3.8  Relationship between First, Second and Third Fundamental Forms . . . . . 96
  3.9  Relationship between Surface Basis Vectors and their Derivatives . . . . . 97
       3.9.1  Codazzi-Mainardi Equations . . . . . . . . . . . . . . . . . 100
  3.10 Sphere Mapping . . . . . . . . . . . . . . . . . . . . . . . . . . . . . 101
  3.11 Global Surface Theorems . . . . . . . . . . . . . . . . . . . . . . . . 102
  3.12 Exercises . . . . . . . . . . . . . . . . . . . . . . . . . . . . . . . . . 102

**4 Curvature**     **110**
  4.1  Curvature Vector . . . . . . . . . . . . . . . . . . . . . . . . . . . . 110
  4.2  Normal Curvature . . . . . . . . . . . . . . . . . . . . . . . . . . . . 113
       4.2.1  Meusnier Theorem . . . . . . . . . . . . . . . . . . . . . . 115
  4.3  Geodesic Curvature . . . . . . . . . . . . . . . . . . . . . . . . . . . 116
  4.4  Principal Curvatures and Directions . . . . . . . . . . . . . . . . . . 118
  4.5  Gaussian Curvature . . . . . . . . . . . . . . . . . . . . . . . . . . . 124
  4.6  Mean Curvature . . . . . . . . . . . . . . . . . . . . . . . . . . . . . 131
  4.7  Theorema Egregium . . . . . . . . . . . . . . . . . . . . . . . . . . . 132
  4.8  Gauss-Bonnet Theorem . . . . . . . . . . . . . . . . . . . . . . . . . 133
  4.9  Local Shape of Surface . . . . . . . . . . . . . . . . . . . . . . . . . 137
  4.10 Umbilical Point . . . . . . . . . . . . . . . . . . . . . . . . . . . . . 141
  4.11 Exercises . . . . . . . . . . . . . . . . . . . . . . . . . . . . . . . . . 142

**5 Special Curves**     **150**
  5.1  Straight Line . . . . . . . . . . . . . . . . . . . . . . . . . . . . . . . 150
  5.2  Plane Curve . . . . . . . . . . . . . . . . . . . . . . . . . . . . . . . 150
  5.3  Involute and Evolute . . . . . . . . . . . . . . . . . . . . . . . . . . . 151
  5.4  Bertrand Curve . . . . . . . . . . . . . . . . . . . . . . . . . . . . . 151
  5.5  Spherical Indicatrix . . . . . . . . . . . . . . . . . . . . . . . . . . . 153
  5.6  Spherical Curve . . . . . . . . . . . . . . . . . . . . . . . . . . . . . 154
  5.7  Geodesic Curve . . . . . . . . . . . . . . . . . . . . . . . . . . . . . 154
  5.8  Line of Curvature . . . . . . . . . . . . . . . . . . . . . . . . . . . . 161
  5.9  Asymptotic Line . . . . . . . . . . . . . . . . . . . . . . . . . . . . . 163
  5.10 Conjugate Direction . . . . . . . . . . . . . . . . . . . . . . . . . . . 166
  5.11 Exercises . . . . . . . . . . . . . . . . . . . . . . . . . . . . . . . . . 167

## 6 Special Surfaces — 172
- 6.1 Plane Surface — 172
- 6.2 Quadratic Surface — 172
- 6.3 Ruled Surface — 173
- 6.4 Developable Surface — 175
- 6.5 Isometric Surface — 176
- 6.6 Tangent Surface — 177
- 6.7 Minimal Surface — 178
- 6.8 Exercises — 179

## 7 Tensor Differentiation — 181
- 7.1 Exercises — 185

## References — 187

## Index — 188

# Nomenclature

In the following table, we define some of the common symbols, notations and abbreviations which are used in the book to provide easy access to the reader.

| | |
|---|---|
| $\nabla$ | nabla differential operator |
| $\nabla^2$ | Laplacian operator |
| $\sim$ | isometric to |
| , subscript | partial derivative with respect to the following index(es) |
| ; subscript | covariant derivative with respect to the following index(es) |
| 1D, 2D, 3D, $n$D | one-dimensional, two-dimensional, three-dimensional, $n$-dimensional |
| overdot (e.g. $\dot{\mathbf{r}}$) | derivative with respect to general parameter $t$ |
| prime (e.g. $\mathbf{r}'$) | derivative with respect to natural parameter $s$ |
| $\delta/\delta t$ | absolute derivative with respect to $t$ |
| $\partial_\alpha, \partial_i$ | partial derivative with respect to $\alpha^{th}$ and $i^{th}$ variables |
| $a$ | determinant of surface covariant metric tensor |
| $\mathbf{a}$ | surface covariant metric tensor |
| $a_{11}, a_{12}, a_{22}$ | coefficients of surface covariant metric tensor |
| $a^{11}, a^{12}, a^{22}$ | coefficients of surface contravariant metric tensor |
| $a_{\alpha\beta}, a^{\alpha\beta}, a^\beta_\alpha$ | surface metric tensor or its components |
| $b$ | determinant of surface covariant curvature tensor |
| $\mathbf{b}$ | surface covariant curvature tensor |
| $\mathbf{B}$ | binormal unit vector of space curve |
| $b_{11}, b_{12}, b_{22}$ | coefficients of surface covariant curvature tensor |
| $b_{\alpha\beta}, b^{\alpha\beta}, b^\beta_\alpha$ | surface curvature tensor or its components |
| $C$ | curve |
| $\bar{C}_\mathbf{B}, \bar{C}_\mathbf{N}, \bar{C}_\mathbf{T}$ | spherical indicatrices of curve $C$ |
| $C_e, C_i$ | evolute and involute curves |
| $C^n$ | of class $n$ |
| $c_{\alpha\beta}, c^{\alpha\beta}, c^\beta_\alpha$ | tensor of third fundamental form or its components |
| $\mathbf{d}$ | Darboux vector |
| $\mathbf{d}_1, \mathbf{d}_2$ | unit vectors in Darboux frame |
| det | determinant of matrix |
| $ds$ | length of infinitesimal element of curve |
| $ds_\mathbf{B}, ds_\mathbf{N}, ds_\mathbf{T}$ | length of line element in binormal, normal and tangent directions |
| $d\sigma$ | area of infinitesimal element of surface |
| $e, f, g$ | coefficients of second fundamental form |
| $E, F, G$ | coefficients of first fundamental form |
| $\mathcal{E}, \mathcal{F}, \mathcal{V}$ | number of edges, faces and vertices of polyhedron |
| $\mathbf{E}_i, \mathbf{E}^j$ | covariant and contravariant space basis vectors |
| $\mathbf{E}_\alpha, \mathbf{E}^\beta$ | covariant and contravariant surface basis vectors |
| Eq./Eqs. | Equation/Equations |

| | |
|---|---|
| $f$ | function |
| Fig./Figs. | Figure/Figures |
| $\mathfrak{g}$ | topological genus of closed surface |
| $g_{ij}$, $g^{ij}$ | space metric tensor or its components |
| $H$ | mean curvature |
| $I_S$, $II_S$, $III_S$ | first, second and third fundamental forms |
| $\mathbf{I}_S$, $\mathbf{II}_S$ | tensors of first and second fundamental forms |
| iff | if and only if |
| $J$ | Jacobian of transformation between two coordinate systems |
| $\mathbf{J}$ | Jacobian matrix |
| $K$ | Gaussian curvature |
| $K_t$ | total curvature |
| $L$ | length of curve |
| $\mathbf{n}$ | normal unit vector to surface |
| $\mathbf{N}$ | principal normal unit vector to curve |
| $P$ | point |
| $r$, $R$ | radius |
| $\mathcal{R}$ | Ricci curvature scalar |
| $\mathbf{r}$ | position vector |
| $\mathbf{r}_\alpha$, $\mathbf{r}_{\alpha\beta}$ | $1^{st}$ and $2^{nd}$ partial derivative of $\mathbf{r}$ with respect to subscripted variables |
| $R_1$, $R_2$ | principal radii of curvature |
| $\mathbb{R}^n$ | $n$-dimensional space (usually Euclidean) |
| $R_{ij}$, $R^i_j$ | Ricci curvature tensor of $1^{st}$ and $2^{nd}$ kind for space |
| $R_{\alpha\beta}$, $R^\alpha_\beta$ | Ricci curvature tensor of $1^{st}$ and $2^{nd}$ kind for surface |
| $R_{ijkl}$ | Riemann-Christoffel curvature tensor of $1^{st}$ kind for space |
| $R_{\alpha\beta\gamma\delta}$ | Riemann-Christoffel curvature tensor of $1^{st}$ kind for surface |
| $R^i_{jkl}$ | Riemann-Christoffel curvature tensor of $2^{nd}$ kind for space |
| $R^\alpha_{\beta\gamma\delta}$ | Riemann-Christoffel curvature tensor of $2^{nd}$ kind for surface |
| $R_\kappa$ | radius of curvature |
| $R_\tau$ | radius of torsion |
| $s$ | natural parameter of curve representing arc length |
| $S$ | surface |
| $S_T$ | tangent surface of space curve |
| $t$ | general parameter of curve |
| $T$ | function period |
| $\mathbf{T}$ | tangent unit vector of space curve |
| $T_P S$ | tangent space of surface $S$ at point $P$ |
| tr | trace of matrix |
| $\mathbf{u}$ | geodesic normal vector |
| $u^1$, $u^2$ | surface coordinates |
| $u^\alpha$ | surface coordinate |
| $u$, $v$ | surface coordinates |
| $x^i$ | space coordinate |

| | |
|---|---|
| $x^i_\alpha$ | surface basis vector in full tensor notation |
| $x, y, z$ | coordinates in 3D space (usually Cartesian) |
| $[ij, k]$ | Christoffel symbol of $1^{st}$ kind for space |
| $[\alpha\beta, \gamma]$ | Christoffel symbol of $1^{st}$ kind for surface |
| $\Gamma^k_{ij}$ | Christoffel symbol of $2^{nd}$ kind for space |
| $\Gamma^\gamma_{\alpha\beta}$ | Christoffel symbol of $2^{nd}$ kind for surface |
| $\delta_{ij}, \delta^{ij}, \delta^j_i$ | covariant, contravariant and mixed Kronecker delta |
| $\delta^{ij}_{kl}$ | generalized Kronecker delta |
| $\Delta$ | discriminant of quadratic equation |
| $\epsilon_{i_1...i_n}, \epsilon^{i_1...i_n}$ | covariant and contravariant relative permutation tensor in $n$D space |
| $\underline{\epsilon}_{i_1...i_n}, \underline{\epsilon}^{i_1...i_n}$ | covariant and contravariant absolute permutation tensor in $n$D space |
| $\theta$ | angle or parameter |
| $\theta_s$ | sum of interior angles of polygon |
| $\kappa$ | curvature of curve |
| $\kappa_1, \kappa_2$ | principal curvatures of surface at a given point |
| $\kappa_\mathbf{B}, \kappa_\mathbf{T}$ | curvature of binormal and tangent spherical indicatrices |
| $\kappa_g, \kappa_n$ | geodesic and normal curvatures |
| $\kappa_{gu}, \kappa_{gv}$ | geodesic curvatures of $u$ and $v$ coordinate curves |
| $\kappa_{nu}, \kappa_{nv}$ | normal curvatures of $u$ and $v$ coordinate curves |
| $\mathbf{K}$ | curvature vector |
| $\mathbf{K}_g, \mathbf{K}_n$ | geodesic and normal components of curvature vector |
| $\lambda$ | direction parameter of surface |
| $\xi$ | real parameter |
| $\rho$ | pseudo-radius of pseudo-sphere |
| $\rho, \phi$ | polar coordinates of plane |
| $\rho, \phi, z$ | cylindrical coordinates of 3D space |
| $\sigma$ | area of surface patch |
| $\tau$ | torsion of curve |
| $\tau_\mathbf{B}, \tau_\mathbf{T}$ | torsion of binormal and tangent spherical indicatrices |
| $\tau_g$ | geodesic torsion |
| $\phi$ | angle or parameter |
| $\chi$ | Euler characteristic |
| $\omega$ | real parameter |

# Chapter 1
# Preliminaries

In this chapter, we provide preliminary materials in the form of a general introduction about differential geometry, remarks about the conventions and notations used in this book, classification of the properties of curves and surfaces, and a general mathematical background related to differential geometry of curves and surfaces.

## 1.1 Differential Geometry

Differential geometry is a branch of mathematics that largely employs methods and techniques of other branches of mathematics such as differential and integral calculus, topology and tensor analysis to investigate geometrical issues related to abstract objects, such as space curves and surfaces, and their properties where these investigations are mostly focused on these properties at small scales. The investigations of differential geometry also include characterizing categories of these objects. There is also a close link between differential geometry and the disciplines of differential topology and differential equations. Differential geometry may be contrasted with "algebraic geometry" which is another branch of geometry that uses algebraic tools to investigate geometric issues mainly of global nature.

The investigation of the properties of curves and surfaces in differential geometry are closely linked. For instance, investigating the characteristics of space curves is extensively exploited in the investigation of surfaces since common properties of surfaces are defined and quantified in terms of the properties of curves embedded in these surfaces. For example, several aspects of the surface curvature at a point are defined and quantified in terms of the parameters of the surface curves passing through that point.

## 1.2 General Remarks, Conventions and Notations

First, we should remark that the present book is largely based on investigating curves and surfaces embedded in a 3D flat space coordinated by a rectangular Cartesian system. In most cases, "surface" and "space" in the present book mean 2D and 3D manifolds respectively.

Another remark is that twisted curves can reside in a 2D manifold (surface) or in a higher dimensionality manifold (usually 3D space). Hence we usually use "surface curves" and "space curves" to refer to the type of the manifold of residence. However, in most cases a single curve can be viewed as resident of more than one manifold and hence it is a surface and space curve at the same time. For example, a curve embedded in a surface which in its turn is embedded in a 3D space is a surface curve and a space curve at the same time. Consequently, in this book these terms should be interpreted flexibly. Many statements formulated in terms of a particular type of manifold can be correctly and easily extended to another type with minimal adjustments of dimensionality and symbolism. Moreover, "space" in some statements should be understood in its general meaning as a manifold embracing the curve not as opposite to "surface" and hence it can include a 2D space, i.e. surface.

Following the convention of several authors, when discussing issues related to 2D and 3D manifolds the Greek indices range over 1, 2 while the Latin indices range over 1, 2, 3. Therefore, the use of Greek and Latin indices should in general indicate the type of the intended manifold unless it is stated otherwise. We use $u$ indexed with superscript Greek letters (e.g. $u^\alpha$ and $u^\beta$) to symbolize surface coordinates,[1] while we largely use $x$ indexed with superscript Latin letters (e.g. $x^i$ and $x^j$) to represent space Cartesian coordinates although they are sometimes used to represent space general curvilinear coordinates. Comments are usually added to clarify the situation if necessary.

A related issue is that the indexed $\mathbf{E}$ are mostly used for the surface, rather than the space, basis vectors where they are subscripted or superscripted with Greek indices, e.g. $\mathbf{E}_\alpha$ and $\mathbf{E}^\beta$. However, in some cases indexed $\mathbf{E}$ are used for space basis vectors; in which case they are distinguished by using Latin indices, e.g. $\mathbf{E}_i$ and $\mathbf{E}^j$. When the basis vectors are indexed numerically rather than symbolically, the distinction between surface and space bases should be obvious from the context if it is not stated explicitly.

Regarding the Christoffel symbols of the first and second kind of various manifolds, they may be based on the space metric or the surface metric. Hence, when a number of Christoffel symbols in a certain context or equation are based on more than one metric, the type of indices, i.e. Greek or Latin, can be used as an indicator to the underlying metric where the Greek indices represent the surface (e.g. $[\alpha\beta, \gamma]$ and $\Gamma^\gamma_{\alpha\beta}$) while the Latin indices represent the space (e.g. $[ij, k]$ and $\Gamma^k_{ij}$). Nevertheless, comments are generally added to account for potential oversight. In particular, when the Christoffel symbols are numbered (e.g. $\Gamma^1_{22}$), instead of being indexed symbolically, comments will be added to clarify the situation.

For brevity, convenience and clean notation in certain contexts, we use an overdot (e.g. $\dot{\mathbf{r}}$) to indicate derivative with respect to a general parameter $t$ while we use a prime (e.g. $\mathbf{r}'$) to indicate derivative with respect to a natural parameter $s$ representing arc length. For the same reasons, subscripts are also used occasionally to symbolize partial derivative with respect to the subscript variables, e.g. $f_v$ and $\mathbf{r}_{\alpha\beta}$. We should also remark that we follow the summation convention, which is largely used in tensor calculus. Comments are added in a few exceptional cases where this convention does not apply.

We deliberately use a variety of notations for the same concepts for the purpose of convenience and to familiarize the reader with different notations all of which are in common use in the literature of differential geometry and tensor calculus. Having proficiency in these subjects requires familiarity with these various, and sometimes conflicting, notations. Moreover, in some situations the use of one of these notations or the other is either necessary or advantageous depending on the location and context. An important example

---

[1] The surface coordinates may also be called the Gaussian coordinates.

of using different notations is that related to surface coordinates where we use both $u, v$ and $u^1, u^2$ to represent these coordinates since each one of these notations has advantages over the other depending on the context; moreover the latter is necessary for expressing the equations of differential geometry in indicial tensor forms.

Another important example related to the use of a variety of notations is the employment of different symbols for the coefficients of the first and second fundamental forms (i.e. $E, F, G, e, f, g$) on one hand and the coefficients of the surface covariant metric tensor and the surface covariant curvature tensor (i.e. $a_{11}, a_{12}, a_{22}, b_{11}, b_{12}, b_{22}$) on the other despite the equivalence of these coefficients, that is $(E, F, G, e, f, g) = (a_{11}, a_{12}, a_{22}, b_{11}, b_{12}, b_{22})$, and hence these different notations can be replaced by just one. However, we keep both notations as they are both in common use in the literature of differential geometry and tensor calculus; moreover in many situations the use of one of these notations in particular is either necessary or advantageous, as indicated earlier.

We should also mention that we use bold face symbols (e.g. $\mathbf{A}$) to mark tensors of rank $> 0$ (including matrices which represent such tensors) in their symbolic notation, while we use normal face symbols (e.g. $c$) for labeling scalars as well as tensors of rank $> 0$ and their components in their indicial form where in the latter case the symbols are indexed (e.g. $A^i$ and $b_\alpha^\beta$). Also, square brackets containing arrays are used to represent matrices while vertical bars are used to represent determinants.

## 1.3 Classifying the Properties of Curves and Surfaces

There are two main classifications for the properties of curves and surfaces embedded in higher dimensionality spaces; these classifications are local properties versus global properties, and intrinsic properties versus extrinsic properties. In the following subsections we briefly investigate these overlapping classifications.

### 1.3.1 Local versus Global Properties

The properties of curves and surfaces may be categorized into two main groups: local and global where these properties describe the geometry of the curves and surfaces *in the small* and *in the large* respectively. The local properties correspond to the characteristics of the object in the immediate neighborhood of a point on the object such as the curvature of a curve or surface at that point, while the global properties correspond to the characteristics of the object on a large scale and over extended parts of the object such as the number of stationary points of a curve or a surface or being a one-side surface like Mobius strip (Fig. 1) which is locally a double-side surface.

As indicated earlier, differential geometry of space curves and surfaces is mainly concerned with the local properties. The investigation of global properties normally involve topological treatments which are beyond the common tools and methods that are usually employed in differential geometry. In fact, there is a special branch of differential geometry dedicated to the investigation of global (or *in the large*) properties. Regarding the present book, it is mostly limited to differential geometry *in the small* although a number

*1.3.2 Intrinsic versus Extrinsic Properties*  11

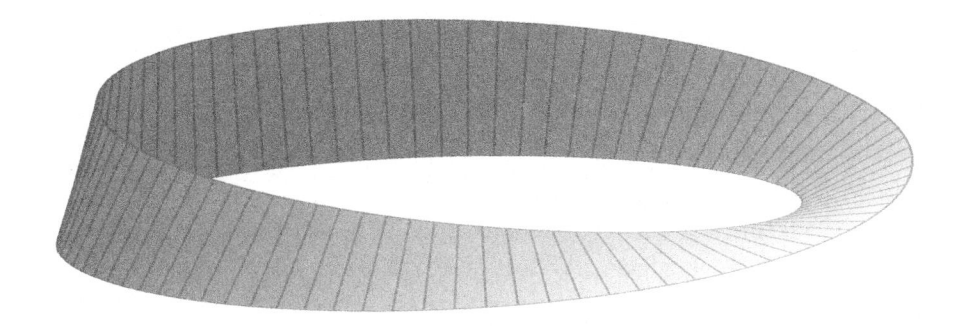

Figure 1: Mobius strip.

of global differential geometric issues are investigated following the tradition pursued in the common textbooks of differential geometry.

## 1.3.2 Intrinsic versus Extrinsic Properties

Another classification of the properties of curves and surfaces, which is based on their relation to the embedding external space which they reside in, may be made where the properties are divided into intrinsic and extrinsic. The first category corresponds to those properties which are independent in their existence and definition from the ambient space which embraces the object such as the distance along a given curve or the Gaussian curvature of a surface at a given point (see § 4.5), while the second category is related to those properties which depend in their existence and definition on the external embedding space such as having a normal vector at a point on the curve or the surface.

More technically, the intrinsic properties[2] are defined and expressed in terms of the metric tensor which is formulated in differential geometry as the first fundamental form (see § 3.5) while the extrinsic properties are expressed in terms of the surface curvature tensor which is formulated in differential geometry as the second fundamental form (see § 3.6). As we will see in several places of this book, some quantities can be expressed once in terms of the coefficients of the first fundamental form exclusively and once in terms of expressions involving the coefficients of the second fundamental form as well. In this regard, a quantity is classified as intrinsic if it can be expressed as a function of the coefficients of the first fundamental form only even if it can also be expressed in terms involving the coefficients of the second fundamental form. We should also remark that extrinsic properties are normally expressed in terms of both forms although these properties are characterized by being expressed necessarily in terms of the second form since this is the feature that distinguishes them from intrinsic properties.

The idea of intrinsic and extrinsic properties may be illustrated by an inhabitant of a surface with a 2D perception (hereafter this creature will be called "2D inhabitant") where he can detect and measure intrinsic properties but not extrinsic properties as the former do not require appealing to an external embedding 3D space in which the surface is immersed

---

[2] The main focus here is the properties of surfaces although some of these qualifications can be extended to surface curves.

while the latter do. Hence, in simple terms all the properties that can be detected and measured by a 2D inhabitant are intrinsic to the surface while all the other properties are extrinsic. A 1D inhabitant of a curve may also be used, to a lesser extent, analogously to distinguish between intrinsic and extrinsic properties of surface and space curves (refer for example to § 2.3).

The so-called "intrinsic geometry" of the surface comprises the collection of all the intrinsic properties of the surface. When two surfaces can have a coordinate system on each such that the first fundamental forms of the two surfaces are identical at each pair of corresponding points on the two surfaces then the two surfaces have identical intrinsic geometry. Such surfaces are isometric (see § 6.5) and can be mapped on each other by a transformation that preserves the curve lengths, the angles and the surface areas. Moreover, if these surfaces are subjected to identical coordinate transformations, they remain identical in their intrinsic properties.

## 1.4 General Mathematical Background

In this section, we provide a general mathematical background outlining concepts and definitions needed in general to understand the forthcoming materials of differential geometry. However, it should be remarked that some of the materials provided in the present section which are related to concepts from other subjects of mathematics, such as calculus and topology, are elementary because of the limits on the text size; moreover the book is not prepared as a text about these subjects. The purpose of these outlines and definitions is to provide a basic understanding of the related ideas. The readers are, therefore, advised to refer to textbooks on those subjects for more technical and extensive revelations. This section also contains some materials that may not be needed in the forthcoming chapters but they are usually discussed and investigated in differential geometry texts. Our objective of including such materials is for the present book to be more comprehensive and compatible with similar differential geometry texts.

### 1.4.1 Geometry and Topology

Here, we define some geometric shapes which are used as examples and prototypes in the forthcoming sections and chapters and they may not be familiar to some readers. Common geometric shapes (like straight line, circle and sphere) are common knowledge at this level and hence they will not be defined. We also introduce some basic topological concepts which will be needed in the forthcoming parts of the book.

A surface of revolution is an axially symmetric surface generated by a plane curve $C$ (see § 5.2) revolving around a straight line $L$ contained in the plane of the curve but not intersecting the curve. The curve $C$ is called the profile of the surface and the line $L$ is called the axis of revolution which is also the axis of symmetry of the surface. Meridians of a surface of revolution are plane curves on the surface formed by the intersection of a plane containing the axis of revolution with the surface, and hence the meridians are identical versions of the profile curve $C$. Parallels of a surface of revolution are circles

## 1.4.1 Geometry and Topology

generated by intersecting the surface by planes perpendicular to the axis of revolution, and hence they represent the paths of specific points on the profile curve $C$. For spheres, these curves are called meridians of longitude and parallels of latitude. On any surface of revolution, meridians and parallels intersect at right angles.

A surface of revolution may be generated in its simplest form by revolving a curve $x = f(z)$ around the $z$-axis of a Cartesian coordinate system. The surface can then be parameterized as:

$$x = f \cos \phi \tag{1}$$
$$y = f \sin \phi \tag{2}$$
$$z = z \tag{3}$$

where $(f, \phi, z)$ represent the standard cylindrical coordinates $(\rho, \phi, z)$.

The **helix** (Fig. 2) is a space curve characterized by having a tangent vector that forms a constant angle with a specified direction which is the direction defined by its axis of rotation. The helix can be defined parametrically by:

$$x = a \cos(\theta) \tag{4}$$
$$y = a \sin(\theta) \tag{5}$$
$$z = b \theta \tag{6}$$

where $a, b$ are non-zero real constants and $\theta$ is a real parameter with $-\infty < \theta < +\infty$. The circle may be regarded as a degenerate form of helix corresponding to $b = 0$, while a straight line may be regarded as another degenerate form of helix corresponding to $a = 0$.

The **torus** (Fig. 3) is a surface of revolution whose profile curve $C$ is a circle. It can be defined parametrically by:

$$x = (R + r \cos \phi) \cos \theta \tag{7}$$
$$y = (R + r \cos \phi) \sin \theta \tag{8}$$
$$z = r \sin \phi \tag{9}$$

where $R$ is the torus radius (i.e. the distance between the center of the generating circle and the center of symmetry of the torus which is the perpendicular projection of the circle center on the axis of revolution), $r$ is the radius of the generating circle ($r < R$), $\phi \in [0, 2\pi)$ is the angle of variation of $r$, and $\theta \in [0, 2\pi)$ is the angle of variation of $R$.

The **ellipsoid** (Fig. 4) is a quadratic surface (see § 6.2) that can be defined parametrically by:

$$x = a \sin \theta \cos \phi \tag{10}$$
$$y = b \sin \theta \sin \phi \tag{11}$$
$$z = c \cos \theta \tag{12}$$

where $a, b, c$ are non-zero real constants and $\theta, \phi$ are real parameters with $0 \leq \theta \leq \pi$ and $0 \leq \phi < 2\pi$.

### 1.4.1 Geometry and Topology

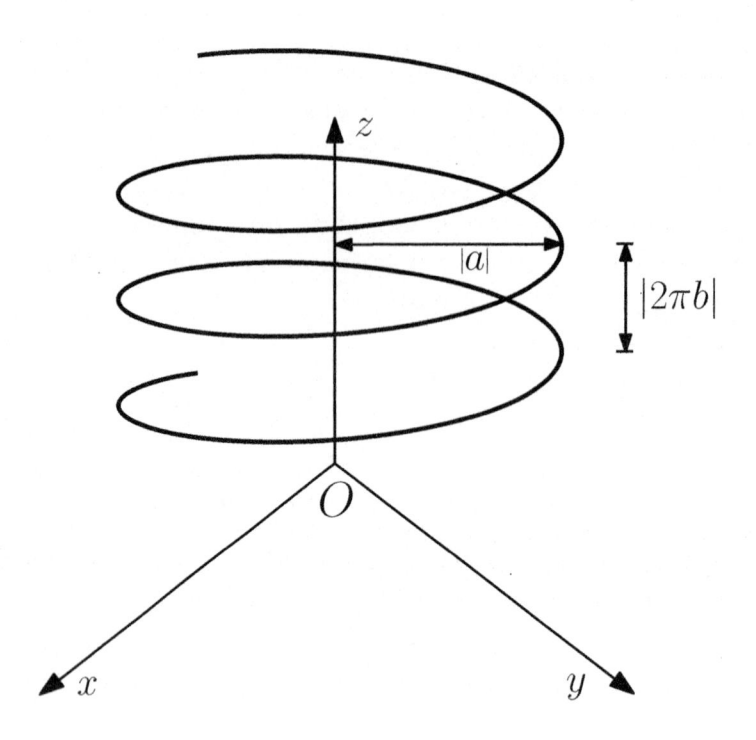

Figure 2: Helix and its parameters.

The **hyperboloid of one sheet** (Fig. 5) is a quadratic surface that can be defined parametrically by:

$$x = a \cosh \xi \, \cos \theta \tag{13}$$
$$y = b \cosh \xi \, \sin \theta \tag{14}$$
$$z = c \sinh \xi \tag{15}$$

where $a, b, c$ are non-zero real constants and $\xi, \theta$ are real parameters with $-\infty < \xi < +\infty$ and $0 \leq \theta < 2\pi$.

The **hyperboloid of two sheets** (Fig. 6) is a quadratic surface that can be defined parametrically by:

$$x = a \sinh \xi \, \cos \theta \tag{16}$$
$$y = b \sinh \xi \, \sin \theta \tag{17}$$
$$z = c \cosh \xi \tag{18}$$

where $a, b, c$ are non-zero real constants and $\xi, \theta$ are real parameters with $0 \leq \xi < \infty$ and $0 \leq \theta < 2\pi$.

The **elliptic paraboloid** (Fig. 7) is a quadratic surface that can be defined parametrically by:

$$x = a \sqrt{\xi} \, \cos \theta \tag{19}$$
$$y = b \sqrt{\xi} \, \sin \theta \tag{20}$$

## 1.4.1 Geometry and Topology

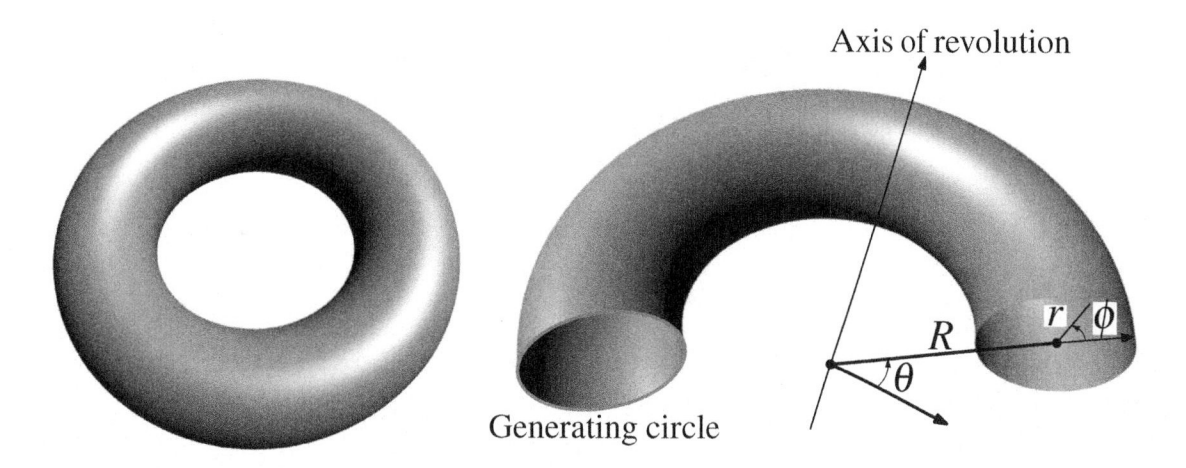

Figure 3: Torus and its parameters.

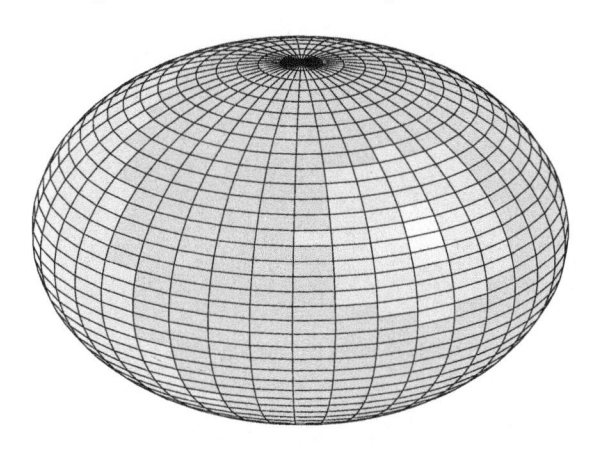

Figure 4: Ellipsoid.

$$z = c\xi \qquad (21)$$

where $a, b, c$ are non-zero real constants and $\xi$, $\theta$ are real parameters with $0 \leq \xi < \infty$ and $0 \leq \theta < 2\pi$.

The **hyperbolic paraboloid** (Fig. 8) is a quadratic surface that can be defined parametrically by:

$$x = a\xi \qquad (22)$$
$$y = b\omega \qquad (23)$$
$$z = c\xi\omega \qquad (24)$$

where $a, b, c$ are non-zero real constants and $\xi$, $\omega$ are real parameters with $-\infty < \xi < +\infty$ and $-\infty < \omega < +\infty$.

The **parabolic cylinder** (Fig. 9) is a quadratic surface generated in its simplest form by translating a parabola (expressed in $y$ as a quadratic function of $x$, or the other way around, in a canonical form) along the $z$-direction. Hence, it can be parameterized by the

### 1.4.1 Geometry and Topology

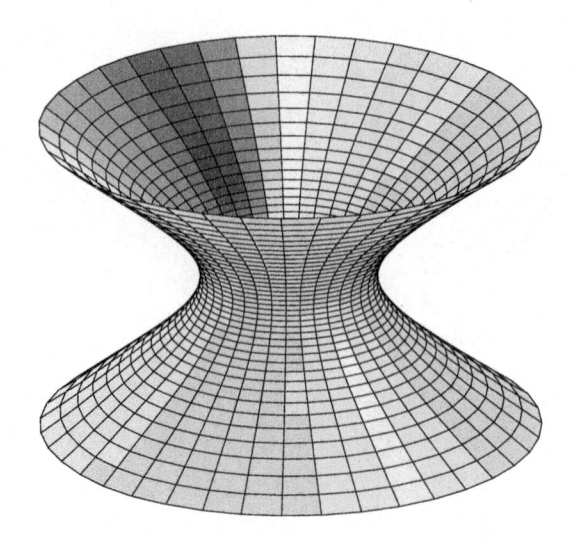

Figure 5: Hyperboloid of one sheet.

following equations:

$$x = \xi \qquad (25)$$
$$y = a\xi^2 \qquad (26)$$
$$z = b\omega \qquad (27)$$

where $a, b$ are non-zero real constants and $\xi, \omega$ are real parameters with $-\infty < \xi < +\infty$ and $-\infty < \omega < +\infty$.

The **catenary** is a plane curve (see § 5.2) with a hyperbolic cosine shape. It can be defined parametrically by:

$$x = a \cosh\left(\frac{\xi}{a}\right) \qquad (28)$$
$$z = \xi \qquad (29)$$

where $a$ is a non-zero real constant and $-\infty < \xi < +\infty$ is a real parameter. The **catenoid** (Fig. 10) is a surface of revolution generated by revolving a catenary around its directrix, which is the $z$-axis in the above formulation, and hence it can be defined parametrically by:

$$x = a \cosh\left(\frac{\xi}{a}\right) \cos\theta \qquad (30)$$
$$y = a \cosh\left(\frac{\xi}{a}\right) \sin\theta \qquad (31)$$
$$z = \xi \qquad (32)$$

where $a$ is a non-zero real constant and $\xi, \theta$ are real parameters with $-\infty < \xi < +\infty$ and $0 \leq \theta < 2\pi$.

## 1.4.1 Geometry and Topology

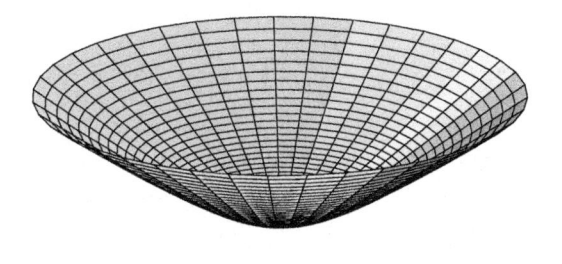

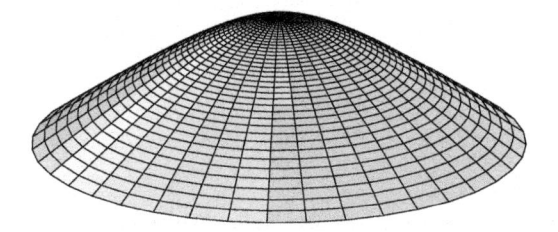

Figure 6: Hyperboloid of two sheets.

The **helicoid** (Fig. 11) is a ruled surface (see § 6.3) with the property that for each point $P$ on the surface there is a helix passing through $P$ and contained entirely in the surface. It can be defined parametrically by:

$$x = a\,\xi \cos\theta \tag{33}$$
$$y = a\,\xi \sin\theta \tag{34}$$
$$z = b\theta \tag{35}$$

where $a, b$ are non-zero real constants and $\xi, \theta$ are real parameters with $-\infty \leq \xi \leq +\infty$ and $-\infty \leq \theta < +\infty$.

The **monkey saddle** (Fig. 12) is a saddle surface that can be defined parametrically by:

$$x = \xi \tag{36}$$
$$y = \omega \tag{37}$$
$$z = \xi^3 - 3\xi\omega^2 \tag{38}$$

where $\xi, \omega$ are real parameters with $-\infty < \xi < +\infty$ and $-\infty < \omega < +\infty$.

The **enneper** (Fig. 13) is a self-intersecting surface that can be defined parametrically by:

$$x = -\frac{\xi^3}{3} + \xi + \xi\omega^2 \tag{39}$$
$$y = -\xi^2\omega - \omega + \frac{\omega^3}{3} \tag{40}$$
$$z = \xi^2 - \omega^2 \tag{41}$$

where $\xi, \omega$ are real parameters with $-\infty < \xi < +\infty$ and $-\infty < \omega < +\infty$.

*1.4.1 Geometry and Topology*                                                                                          18

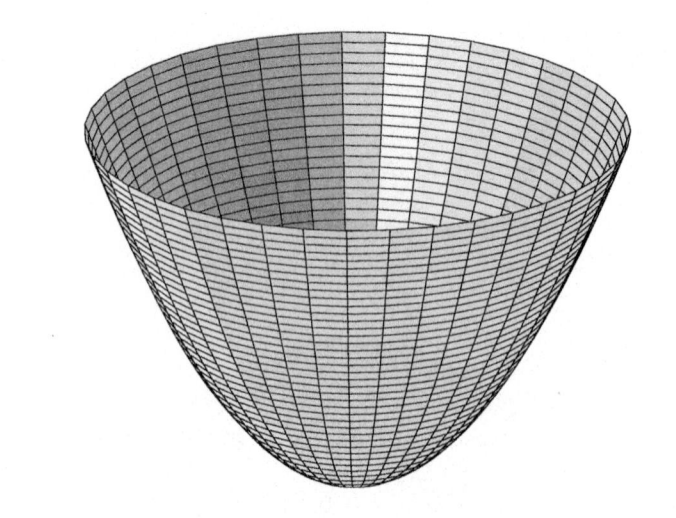

Figure 7: Elliptic paraboloid.

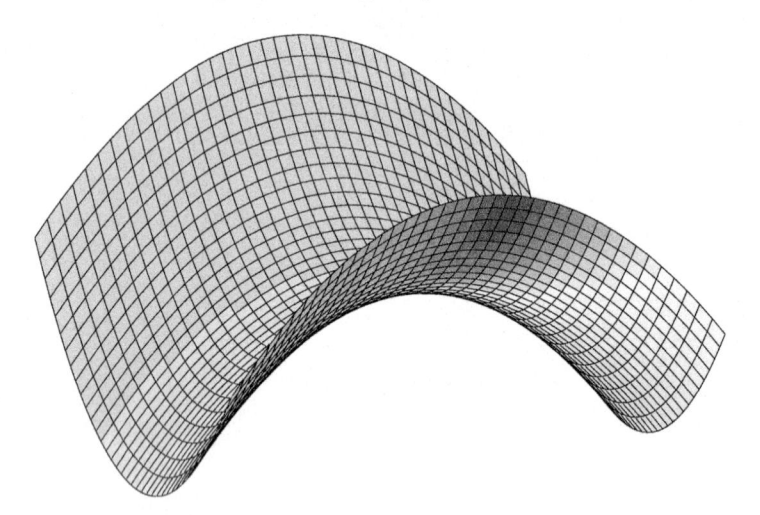

Figure 8: Hyperbolic paraboloid.

The **tractrix** (Fig. 14) is a plane curve starting[3] from the point $(\rho, 0)$ on the $x$-axis with the property that the length of the line segment of its tangent between the tangency point and the point of intersection with the $z$-axis is equal to $\rho$ where $\rho$ is a real constant ($\rho > 0$). Hence, the tractrix is a solution of the following differential equation:

$$\frac{dz}{dx} = \pm \frac{\sqrt{\rho^2 - x^2}}{x} \qquad (0 < x \leq \rho) \qquad (42)$$

with the condition $z(\rho) = 0$. The plus sign in this equation corresponds to the lower part ($z < 0$) while the minus sign corresponds to the upper part. The Beltrami **pseudo-sphere** (Fig. 14) is a surface of revolution generated by revolving a tractrix around its asymptote which is the $z$-axis in the above formulation. The pseudo-sphere can be defined

---

[3] We are assuming the curve is embedded in the $xz$ plane.

## 1.4.1 Geometry and Topology

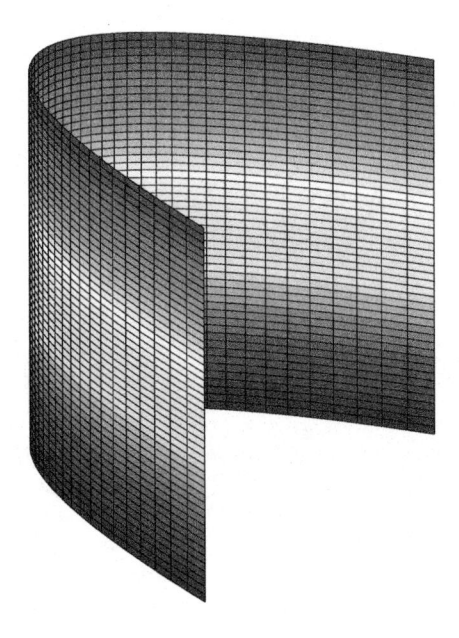

Figure 9: Parabolic cylinder.

parametrically by:

$$x = a \sin\theta \cos\phi \qquad (43)$$
$$y = a \sin\theta \sin\phi \qquad (44)$$
$$z = a \left[\cos\theta + \ln\left(\tan\frac{\theta}{2}\right)\right] \qquad (45)$$

where $a$ is a real constant and $\theta$, $\phi$ are real parameters with $0 < \theta < \pi$ and $0 \leq \phi < 2\pi$. The constant $\rho$ of the tractrix which generates the pseudo-sphere is called the pseudo-radius of the pseudo-sphere.

It should be remarked that the geometric shapes that we defined parametrically in this subsection (e.g. torus, ellipsoid, and hyperbolic paraboloid) can also be defined by other parametric forms as well as by explicit Cartesian and non-Cartesian forms (see e.g. § 6.2).

The **Euler characteristic**, or Euler-Poincare characteristic, is a topological parameter of closed surfaces (see § 3.1) which, for polyhedral surfaces, is given by:

$$\chi = \mathcal{V} + \mathcal{F} - \mathcal{E} \qquad (46)$$

where $\chi$ is the Euler characteristic of the surface, and $\mathcal{V}, \mathcal{F}, \mathcal{E}$ are the numbers of vertices, faces and edges of the polyhedron. Examples of Euler characteristic of some common polyhedrons are given in Fig. 15.

The Euler characteristic can also be defined for more general types of surface. The Euler characteristic of a compact orientable (see § 3.1) non-polyhedral surface, like sphere and torus, can be obtained by polygonal decomposition based on dividing the entire surface into a finite number of non-overlapping curvilinear polygons which share at most edges or

*1.4.2 Functions* 20

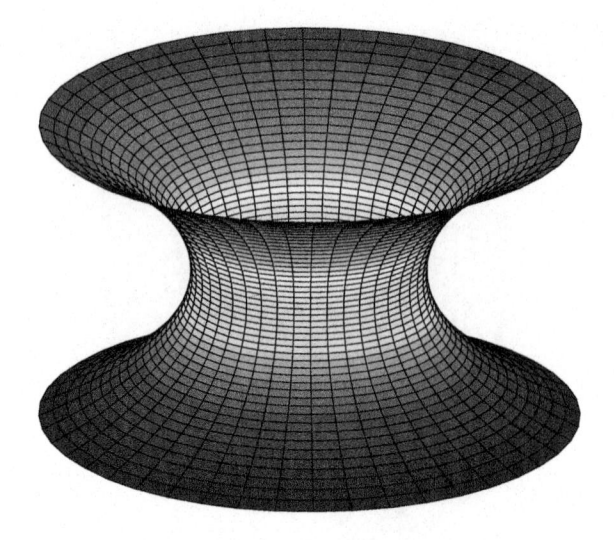

Figure 10: Catenoid.

vertices (see Fig. 16). The above formula (Eq. 46) is then used, as for polyhedral surfaces, to determine the Euler characteristic of the surface.

In simple terms, the topological **genus** of a surface is the number of handles or topological holes on the surface. For example, the ellipsoid (Fig. 4) is a surface of genus 0 while the torus (Fig. 3) is a surface of genus 1. Similarly, the surfaces in Fig. 17 are of genus 2 and 3. For an orientable surface of genus $\mathfrak{g}$ the Euler characteristic $\chi$ is related to the genus by:

$$\chi = 2(1 - \mathfrak{g}) \qquad (47)$$

The length of a straight line segment connecting two points in a Euclidean space is a measure of the distance (in its common sense) between the two points.[4] The length of a polygonal arc (Fig. 18 a) is the sum of the lengths of its straight segments. The length of an arc of a generalized twisted space curve is the limit of the length of an asymptotic polygonal arc (Fig. 18 b) as the length of the longest straight line segment of the asymptotic polygonal arc tends to zero.

The area of a polygonal plane fragment (e.g. triangle or square) is a measure of its two-dimensional expansion.[5] The area of a surface patch consisting entirely of polygonal plane fragments is the sum of the areas of its polygonal plane fragments. The area of a patch of a generalized twisted space surface is the limit of the area of its asymptotic polygonal patch as the area of the largest of its polygonal fragments tends to zero (refer to Fig. 19).

### 1.4.2 Functions

The domain of a functional mapping: $f : \mathbb{R}^m \to \mathbb{R}^n$ is the largest set of $\mathbb{R}^m$ on which the mapping is defined. A bicontinuous function or mapping is a continuous function with a

---
[4] The "length of a straight line segment" may be taken as an axiomatic concept.
[5] The "area of a polygonal plane fragment" may be taken as an axiomatic concept.

*1.4.2 Functions* 21

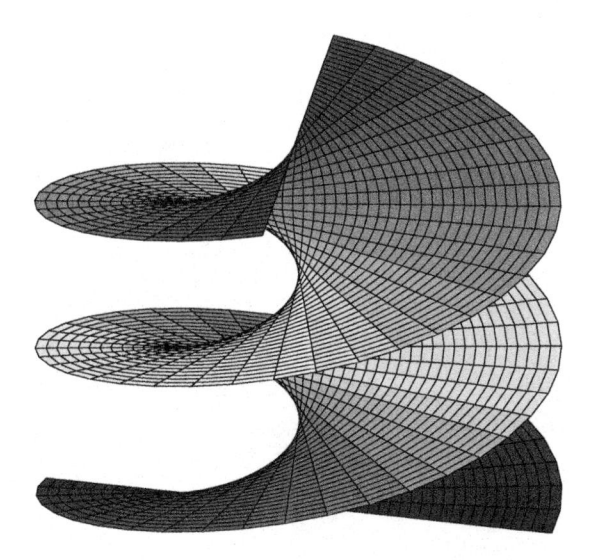

Figure 11: Helicoid.

continuous inverse. A scalar function is of class $C^n$ if the function and all of its first $n$ (but not $n+1$) partial derivatives do exist and are continuous. A vector function (e.g. a position vector representing a space curve or surface) is of class $C^n$ if one of its components is of this class while all the other components are of this class or higher. A curve or a surface is of class $C^n$ if it is mathematically represented by a function of this class.

In this context we should remark that in most cases the purpose of imposing the condition of having a function of class $C^n$ is to have a function which is at least of this class and hence the condition is met by having a function of this class or higher. In fact, this is generally the case in this book when this condition is imposed unless there is an obvious indicator that the function is strictly of this class. Another remark is that there should be no confusion between the bare $C$ symbol and the superscripted $C$ symbol, like $C^2$, as the bare $C$ is usually used in this book to symbolize a curve while the superscripted $C$ stands for the above differentiability condition.

In gross terms, a "smooth" or "sufficiently smooth" curve or surface means that the functional relation that represents the object is sufficiently differentiable for the intended objective, being of class $C^n$ at least where $n$ is the minimum requirement for the differentiability index to satisfy the required conditions.

A deleted neighborhood of a point $P$ on a 1D interval on the real line is defined as the set of all points $x \in \mathbb{R}$ in the interval such that:

$$0 < |x - x_P| < \epsilon \tag{48}$$

where $x_P$ is the coordinate of $P$ on the real line and $\epsilon$ is a positive real number. Hence, the deleted neighborhood includes all the points in the open interval $(x_P - \epsilon, x_P + \epsilon)$ excluding $x_P$ itself. For a space curve (which is not straight in general) represented by $\mathbf{r} = \mathbf{r}(t)$, where $\mathbf{r}$ is the spatial representation of the curve and $t$ is a general parameter in the curve representation, the definition applies to the neighborhood of $t_P$ where $t_P$ is the value of

## 1.4.2 Functions

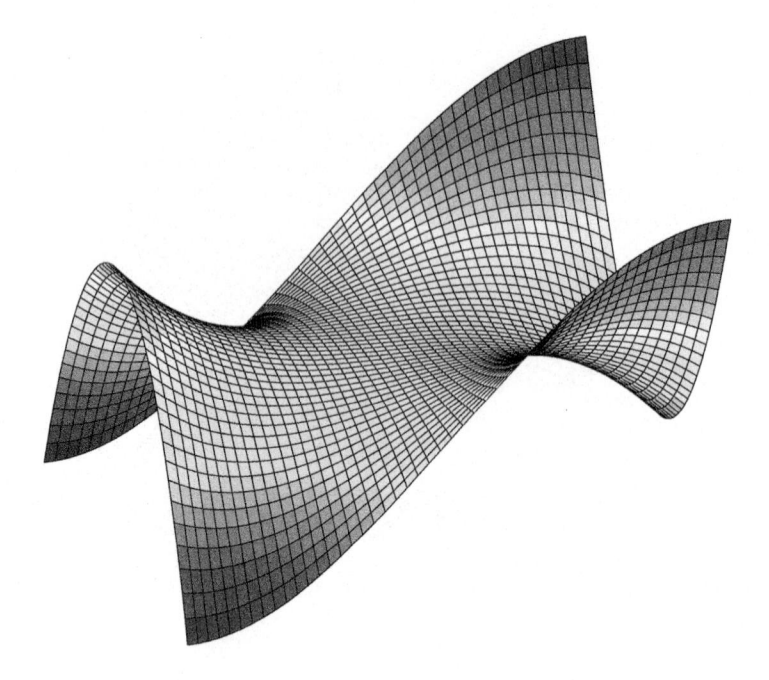

Figure 12: Monkey saddle.

$t$ corresponding to the point $P$ on the curve. Deleted neighborhood of a twisted curve may also be identified by being confined in a circle of radius $\epsilon$ centered at $P$, however this applies only to plane curves. Therefore, it may be identified by being confined in a sphere of radius $\epsilon$ centered at $P$ to be more general.

A deleted neighborhood of a point $P$ on a 2D flat surface is defined as the set of all points $(x, y) \in \mathbb{R}^2$ on the surface such that:

$$0 < \sqrt{(x - x_P)^2 + (y - y_P)^2} < \epsilon \qquad (49)$$

where $(x_P, y_P)$ are the coordinates of $P$ on the plane and $\epsilon$ is a positive real number. Hence, the deleted neighborhood includes all the points inside a circular disc of radius $\epsilon$ and center $(x_P, y_P)$ excluding the center itself. For a space surface (which is not flat in general) represented by $\mathbf{r} = \mathbf{r}(u, v)$, where $\mathbf{r}$ is the spatial representation of the surface and $u, v$ are the surface coordinates on the $uv$ plane (see § 1.4.3) that map on the surface, the definition applies to the neighborhood of $(u_P, v_P)$ where $(u_P, v_P)$ are the coordinates on the 2D $uv$ plane corresponding to the point $P$ on the surface. Similar to twisted curves, deleted neighborhood of a twisted surface may also be identified by being confined in a sphere of radius $\epsilon$ centered at $P$. The above definitions of deleted neighborhoods of curves and surfaces can be easily extended to spaces of higher dimensionality. For instance, in a 3D space the neighborhood is contained in a spherical surface with radius $\epsilon$. However, this is not needed in this book whose focus is space curves and surfaces.

A quadratic expression:

$$Q(x, y) = a_1 x^2 + 2a_2 xy + a_3 y^2 \qquad (50)$$

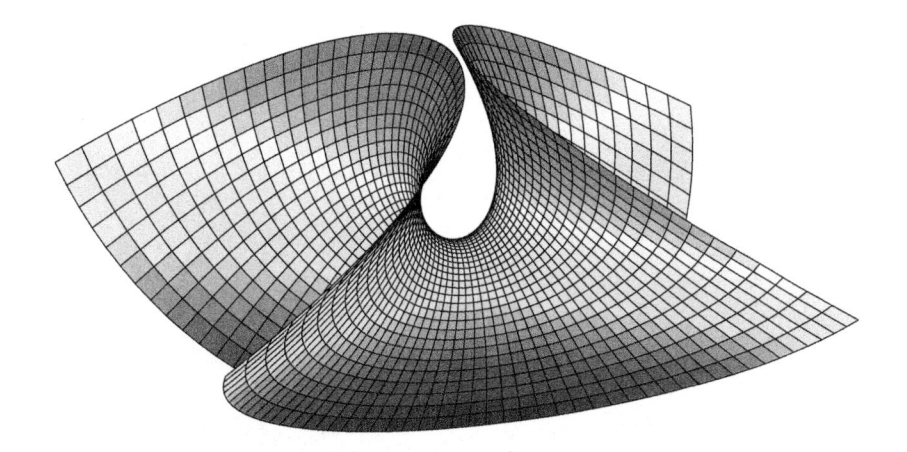

Figure 13: Enneper.

of real coefficients $a_1, a_2, a_3$ and real variables $x, y$ is described as "positive definite" if it possesses positive values ($> 0$) for all pairs $(x, y) \neq (0, 0)$. The sufficient and necessary condition for $Q$ to be positive definite is that:

$$a_1 > 0 \quad \text{and} \quad (a_1 a_3 - a_2 a_2) > 0 \tag{51}$$

These two conditions necessitate a third condition that is: $a_3 > 0$ since these coefficients are real.

The first variation of a functional $F$ may be defined by the Gateaux derivative of the functional as:

$$\delta F(x, h) = \lim_{\xi \to 0} \frac{F(x + \xi h) - F(x)}{\xi} \tag{52}$$

where $x$ and $h$ are variable functions and $\xi$ is a real scalar parameter. The Euler-Lagrange variational principle is a mathematical rule whose objective is to minimize or maximize a certain functional $F(f)$ which depends on a function $f$. It is represented mathematically by a partial differential equation whose solutions optimize the particular functional $F$. The Euler-Lagrange principle in its generic, simple and most common form is given mathematically by:

$$\frac{\partial f}{\partial y} - \frac{d}{dx}\left(\frac{\partial f}{\partial y_x}\right) = 0 \tag{53}$$

where $f(x, y, y_x)$ is a function of the given variables that optimizes the functional $F$, $y$ is a function of $x$, and $y_x$ is the derivative of $y$ with respect to $x$.

## 1.4.3 Coordinates, Transformations and Mappings

The Jacobian $J$ of a transformation between two coordinate systems, labeled as unbarred and barred, is the determinant of the Jacobian matrix **J** of the transformation between

### 1.4.3 Coordinates, Transformations and Mappings

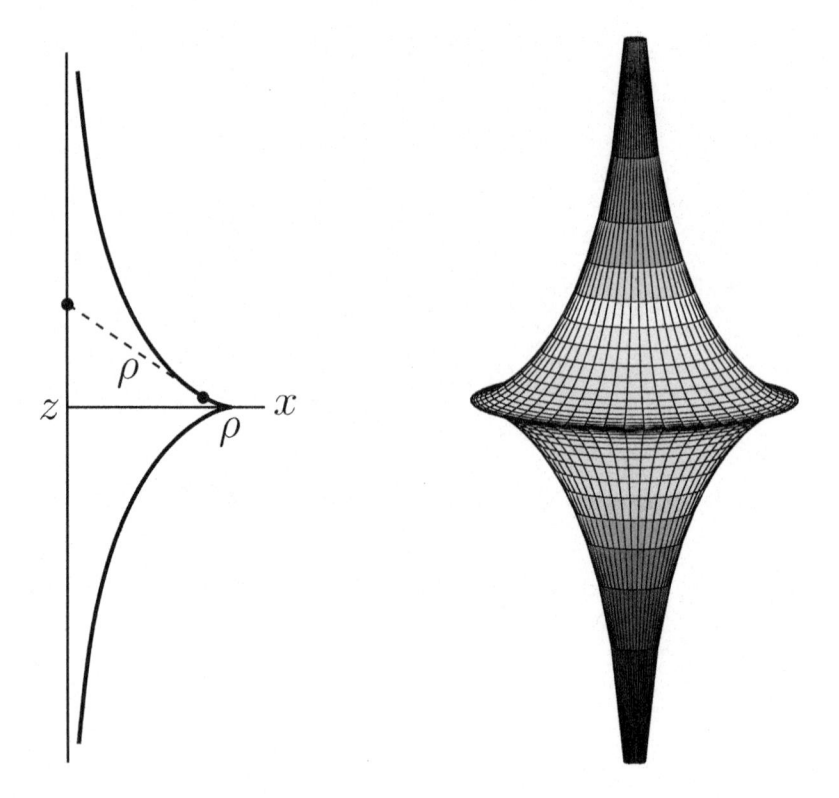

Figure 14: Tractrix (left frame) and pseudo-sphere (right frame).

these systems, that is:

$$J = \det(\mathbf{J}) = \begin{vmatrix} \frac{\partial x^1}{\partial \bar{x}^1} & \frac{\partial x^1}{\partial \bar{x}^2} & \cdots & \frac{\partial x^1}{\partial \bar{x}^n} \\ \frac{\partial x^2}{\partial \bar{x}^1} & \frac{\partial x^2}{\partial \bar{x}^2} & \cdots & \frac{\partial x^2}{\partial \bar{x}^n} \\ \vdots & \vdots & \ddots & \vdots \\ \frac{\partial x^n}{\partial \bar{x}^1} & \frac{\partial x^n}{\partial \bar{x}^2} & \cdots & \frac{\partial x^n}{\partial \bar{x}^n} \end{vmatrix} \tag{54}$$

where the indexed $x$ and $\bar{x}$ are the coordinates in the unbarred and barred coordinate systems in an $n$D space. An admissible coordinate transformation may be defined generically as a mapping represented by a sufficiently differentiable set of equations and it is invertible by having a non-vanishing Jacobian ($J \neq 0$).[6] An invariant property of a curve or a surface is a property which is independent of admissible coordinate transformations and parameterizations. It should be noted that an invariant property may be invariant with respect to certain types of transformation or parameterization but not with respect to others and hence it is sometimes used generically where the context should be taken into consideration for sensible interpretation.

An orthogonal coordinate transformation is a combination of translation, rotation and reflection of axes. The Jacobian of orthogonal transformations is unity, that is $J = \pm 1$. The orthogonal transformation is described as positive *iff* $J = +1$ and negative *iff* $J = -1$.

---

[6] The meaning of "admissible coordinate transformation" may vary depending on the context. We are generally assuming linear transformations.

### 1.4.3 Coordinates, Transformations and Mappings

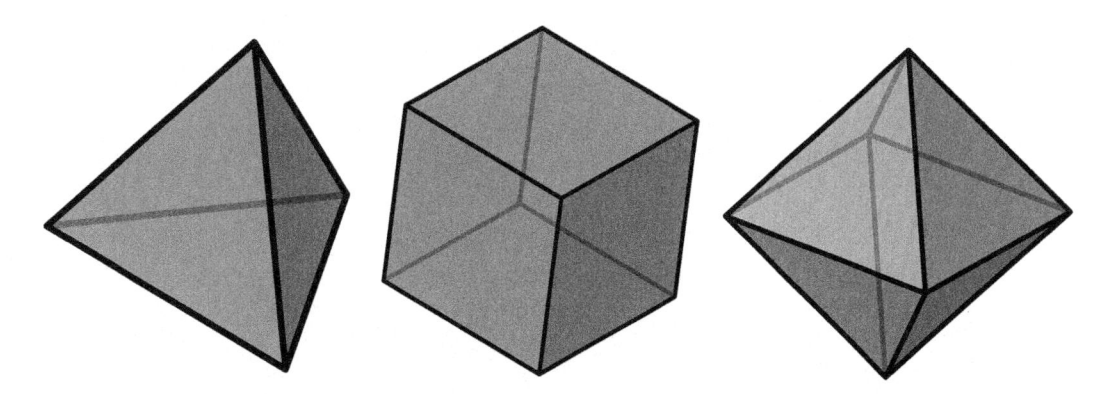

Figure 15: Polyhedral surfaces: tetrahedron (left frame), cube (middle frame) and octahedron (right frame). The Euler characteristic for these surfaces is given respectively by: $\chi = 4 + 4 - 6$, $\chi = 8 + 6 - 12$, and $\chi = 6 + 8 - 12$, which in all cases is equal to 2.

Positive orthogonal transformations consist solely of translation and rotation (possibly trivial ones as in the case of the identity transformation) while negative orthogonal transformations include reflection, by applying an odd number of axes reversal, as well. Positive transformations can be decomposed into an infinite number of continuously varying infinitesimal positive transformations each one of which imitates an identity transformation. Such a decomposition is not possible in the case of negative orthogonal transformations because the shift from the identity transformation to reflection is impossible by a continuous process.

A surface $S$ in a 3D space may be described directly by the three spatial coordinates of a given 3D coordinate system (e.g. $x, y, z$ of a Cartesian system) or by the two surface coordinates (e.g. $u, v$). In the former case, the description is rather familiar where each point on the space surface is identified directly by three spatial coordinates linked by a given functional relation (see § 3.1 and 6.2). In the latter case, a 2D domain over which the surface $S$ is defined is introduced, where this domain is identified by two independent variables (usually $u, v$ or $u^1, u^2$) . The domain is usually assumed, for simplicity, to be a plane (which may be called the $uv$ plane or parameters plane) where a grid of $u$ and $v$ curves are defined over this plane. Again, for simplicity the curves in these $u$ and $v$ families are usually defined as straight lines; moreover, the curves in each family are regularly spaced with the two families of $u$ and $v$ curves being mutually orthogonal (see Fig. 20). Although these assumptions about the domain and about the grid and parameter curves can be violated (e.g. by using a 2D polar or curvilinear coordinate system over the parameters plane instead of the above-described 2D rectangular Cartesian), in most cases there is no advantage in doing so. Along each curve of the $u$ and $v$ families only one variable (either $u$ or $v$) varies while the other variable ($v$ or $u$) is held constant. So, along any curve of the $u$ family $v$ is constant and along any curve of the $v$ family $u$ is constant.

The surface $S$ is then defined by a functional mapping from the $uv$ plane onto the space surface $S$. This functional mapping identifies any point on the surface in the 3D space corresponding to any point in its domain in the $uv$ plane. Hence, the $uv$ grid on

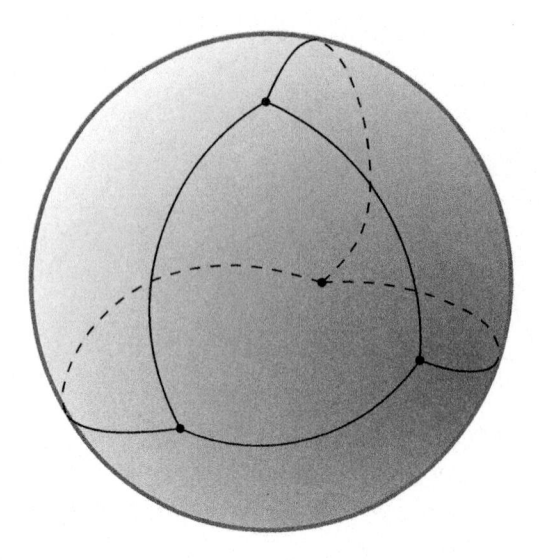

### 1.4.3 Coordinates, Transformations and Mappings

Figure 16: Polygonal decomposition of a sphere into four non-overlapping curvilinear polygons.

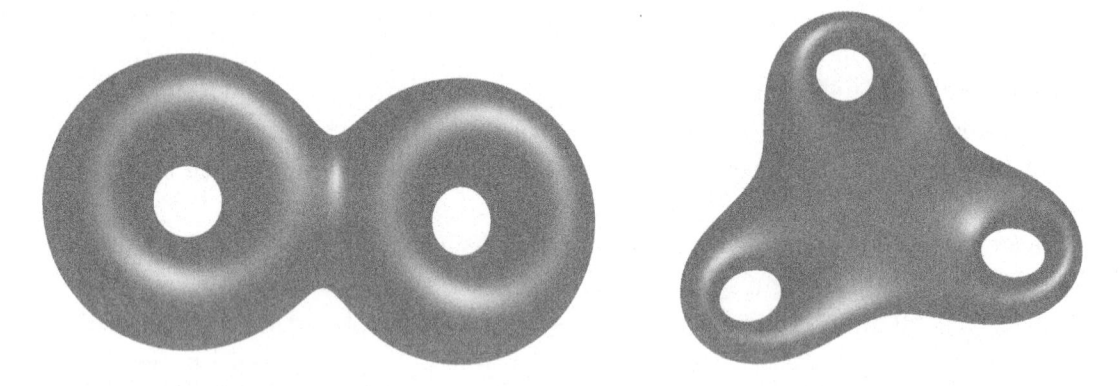

Figure 17: Examples of surfaces of genus 2 (left frame) and genus 3 (right frame).

the $uv$ plane is projected by this functional mapping on a corresponding $uv$ grid on the surface $S$. As we are assuming that the surface is embedded in a 3D space, the functional mapping consists of three independent relations where each relation correlates the $u$ and $v$ coordinates of a given point in the domain to one of the three spatial coordinates $(x^1, x^2, x^3)$ of a corresponding point of the surface in the embedding space, that is: $x^1(u,v)$, $x^2(u,v)$ and $x^3(u,v)$.

While the $uv$ grid on the $uv$ plane is usually a planar, orthogonal and regularly spaced grid, the corresponding grid on the surface $S$ is not necessarily so because it is generally a 3D grid that follows the bends and variations of the twisted space surface. In this context, "coordinate curves" (which are also called parametric curves or parametric lines) on a surface are defined as the map or projection of the $uv$ curves of the $uv$ grid of the parameters plane onto the space surface, as described above, and hence they are curves along which only one coordinate variable ($u$ or $v$) varies while the other coordinate variable

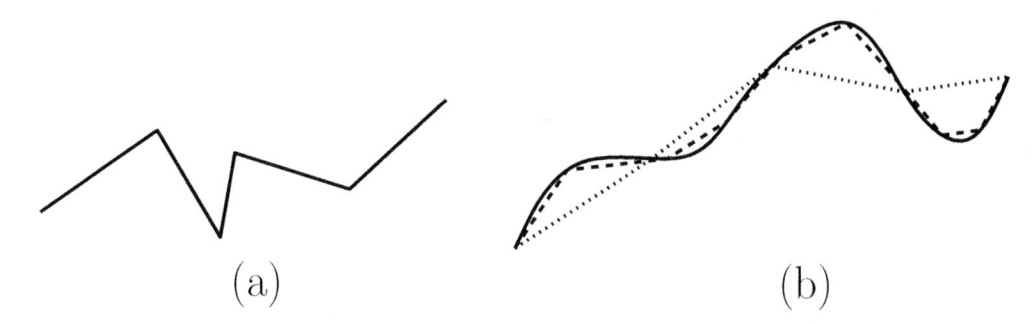

### 1.4.3 Coordinates, Transformations and Mappings

(a)          (b)

Figure 18: (a) Polygonal arc and (b) twisted curve (solid) with two of its asymptotic polygonal arcs (dashed and dotted) where the dashed represents a better approximation to the length of the twisted curve than the dotted.

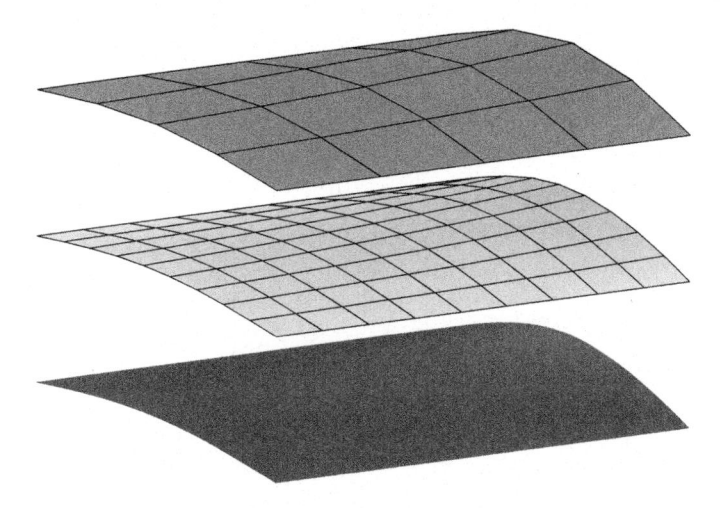

Figure 19: A smooth surface (bottom) approximated by a coarse asymptotic polygonal surface (top) and a fine asymptotic polygonal surface (middle) where the fine represents a better approximation to the area of the smooth surface than the coarse.

($v$ or $u$) remains constant. So, along the $u$ coordinate curves $v$ is held constant while along the $v$ coordinate curves $u$ is held constant (see Fig. 20).

It is noteworthy that although the surface coordinates $u, v$ (or $u^1, u^2$) primarily represent the coordinates on the $uv$ parameters plane which maps on the surface, in the literature of differential geometry they are sometimes used (to ease the notation) as labels for the curves on the surface. Hence, vigilance is required to avoid confusion.

A regular representation of class $C^m$ ($m > 0$) of a surface patch $S$ in a 3D Euclidean space is defined as a functional mapping of an open set $\Omega$ in the $uv$ plane onto $S$ that satisfies the following two conditions:

1. The functional mapping relation is of class $C^m$ over the entire $\Omega$.
2. The Jacobian matrix of the transformation between the representation of the surface in the 3D space and its 2D domain is of rank 2 for all the points in $\Omega$.

We remark that for a functional mapping of the form $\mathbf{S}(u,v) = (S_1(u,v), S_2(u,v), S_3(u,v))$,

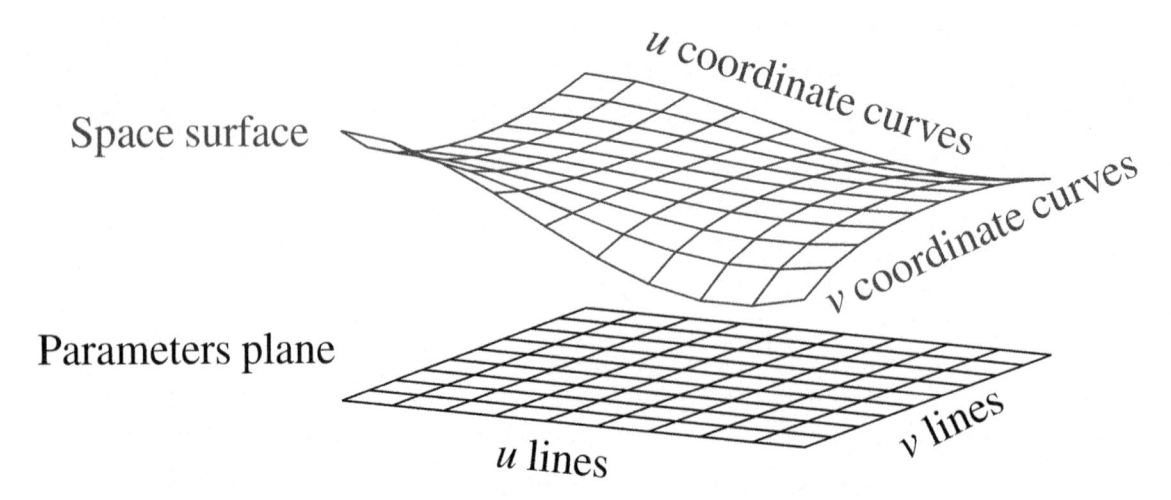

### 1.4.3 Coordinates, Transformations and Mappings

Figure 20: The $uv$ parameters plane corresponding to a space surface and the mapping of the $uv$ grid.

the aforementioned Jacobian matrix is given by:

$$\mathbf{J} = \begin{bmatrix} \partial_u S_1 & \partial_v S_1 \\ \partial_u S_2 & \partial_v S_2 \\ \partial_u S_3 & \partial_v S_3 \end{bmatrix} \tag{55}$$

In fact, having a Jacobian matrix of rank 2 for the transformation, according to the above condition, is equivalent to the condition that $\mathbf{E}_1 \times \mathbf{E}_2 \neq \mathbf{0}$ where $\mathbf{E}_1 = \partial_u \mathbf{r}$ and $\mathbf{E}_2 = \partial_v \mathbf{r}$ are the surface basis vectors (see § 1.4.5, 3.1 and 3.2), which are the tangents to the $u$ and $v$ coordinate curves respectively, and $\mathbf{r} = \mathbf{r}(u,v)$ is the 3D spatial representation of the curves. Having a Jacobian matrix of rank 2 is also equivalent to having a well-defined tangent plane to the surface at the related point. In this regard, we note that "rank" here refers to its meaning in linear algebra and should not be confused with the rank of a tensor which is related to the number of its free indices although there is a connection between the two.

As we will see, a regular point on a surface is a point that satisfies the above condition about the basis vectors. A point on a surface which is not regular is called singular. Singularity occurs either because of a geometric reason, which is the case for instance for the apex of a cone, or because of the particular parametric representation of the surface. While the first type of singularity is inherent and hence it cannot be removed, the second type can be removed by changing the representation.

Corresponding points on two curves refer to two points, one on each curve, with a common value of a common parameter of the two curves. When the two curves have two different parameterizations then a one-to-one correspondence between the two parameters should be established and the corresponding points then refer to two points with corresponding values of the two parameters. Corresponding points on two surfaces can be defined in a similar manner taking into account that surfaces require two independent parameters in their identification.

In many cases of theoretical and practical significance, a mixed tensor $A^i_\alpha$, which is contravariant with respect to transformations in space coordinates $x^i$ and covariant with respect to transformations in surface coordinates $u^\alpha$, may be defined. Following a coordinate transformation in which both the space and surface coordinates change, the tensor $A^i_\alpha$ will be given in the new (barred) system by:

$$\bar{A}^i_\alpha = A^j_\beta \frac{\partial \bar{x}^i}{\partial x^j} \frac{\partial u^\beta}{\partial \bar{u}^\alpha} \tag{56}$$

More generally, tensors with space and surface contravariant indices and space and surface covariant indices (e.g. $A^{i\alpha}_{j\beta}$) can also be defined in a similar manner. The extension of the above transformation rule to include such tensors can be easily achieved by following the obvious pattern seen in the last equation.

### 1.4.4 Intrinsic Distance

The intrinsic distance between two points on a surface is the greatest lower bound (or infimum) of the lengths of all regular arcs connecting the two points on the surface. The intrinsic distance is an intrinsic property of the surface. The intrinsic distance $d$ between two points is invariant under a local isometric mapping (see § 3.1), that is:

$$d(f(P_1), f(P_2)) = d(P_1, P_2) \tag{57}$$

where $f$ is an isometric mapping from a surface $S_1$ to a surface $S_2$ (see § 3.1 and 6.5), $P_1$ and $P_2$ are the two points on $S_1$ and $f(P_1)$ and $f(P_2)$ are their images on $S_2$. In fact, this may be taken as the definition of isometric mapping, i.e. it is the mapping that preserves intrinsic distance.

The following conditions apply to the intrinsic distance $d$ between points $P_1$, $P_2$ and $P_3$:
1. Symmetry: $d(P_1, P_2) = d(P_2, P_1)$.
2. Triangle inequality: $d(P_1, P_3) \leq d(P_1, P_2) + d(P_2, P_3)$.
3. Positive definiteness: $d(P_1, P_2) > 0$ with $d(P_1, P_2) = 0$ *iff* $P_1$ and $P_2$ are the same point.

An arc $C$ connecting two points, $P_1$ and $P_2$, on a surface is described as an arc of minimum length between $P_1$ and $P_2$ if the length of $C$ is equal to the intrinsic distance between $P_1$ and $P_2$. The existence and uniqueness of an arc of minimum length between two specific points on a surface is not guaranteed, i.e. it may not exit and if it does exist it may not be unique (refer to § 5.7 for examples). Yes, for certain types of surface such an arc does exist and it is unique. For example, on a simply connected (see § 3.1) plane surface there exists an arc of minimum length between any two points on the plane and it is unique; this arc is the straight line segment connecting the two points.

### 1.4.5 Basis Vectors

The set of basis vectors in a given manifold plays a pivotal role in the theoretical construction of the geometry of the manifold, and this applies to the basis vectors in differential geometry where these vectors are used in the definition and construction of essential concepts and objects such as the metric tensor of the surface. The set of basis vectors may also be employed to serve as a moving coordinate frame for the enveloping space of their underlying constructions (see § 2.2 and § 3.2).

The differential geometry of curves and surfaces employs two main sets of basis vectors:

1. One set is constructed on space curves and consists of three unit vectors: the tangent **T**, the normal **N** and the binormal **B** to the curve.
2. Another set is constructed on surfaces and consists of two linearly independent vectors which are the tangents to the coordinate curves of the surface, $\mathbf{E}_1 = \frac{\partial \mathbf{r}}{\partial u^1}$ and $\mathbf{E}_2 = \frac{\partial \mathbf{r}}{\partial u^2}$, plus the unit normal vector to the surface **n**, where $\mathbf{r}(u^1, u^2)$ is the spatial representation of a surface coordinate curve, and $u^1$ and $u^2$ are the surface coordinates as discussed earlier and as will be investigated further later in the book (refer to § 1.4.3 and **3.2**).

Each one of the above basis sets is defined on each regular point of the curve or the surface and hence in general the vectors in each one of these basis sets vary from one point to another, i.e. they are position dependent. The vectors $\mathbf{E}_1$ and $\mathbf{E}_2$, which are not necessarily of unit length, vary in magnitude and direction while the unit vectors (which are the rest) vary in direction. Also, while the vectors of the **T, N, B** set are mutually orthogonal, the vectors of the $\mathbf{E}_1, \mathbf{E}_2, \mathbf{n}$ set is not necessarily so since $\mathbf{E}_1$ and $\mathbf{E}_2$ are not orthogonal in general although **n** is orthogonal to both $\mathbf{E}_1$ and $\mathbf{E}_2$.

The surface basis vectors, $\mathbf{E}_1$ and $\mathbf{E}_2$, are given in full tensor notation by $\frac{\partial x^i}{\partial u^\alpha}$ ($i = 1, 2, 3$ and $\alpha = 1, 2$) which is usually abbreviated as $x^i_\alpha$. These vectors can be seen as contravariant space vectors or as covariant surface vectors. Further details about this will be given in § 3.3. We remark that other sets of basis vectors are also defined and employed in differential geometry, as we will see in the future (refer to § 4.1 and 4.4).

## 1.4.6 Flat and Curved Spaces

A manifold, such as a 2D surface or a 3D space, is called "flat" if it is possible to find a coordinate system for the manifold with a diagonal metric tensor whose all diagonal elements are ±1; the space is called "curved" otherwise. More formally, an $n$D space is described as a flat space *iff* it is possible to find a coordinate system for which the line element $ds$ is given by:

$$(ds)^2 = \zeta_1 (dx^1)^2 + \zeta_2 (dx^2)^2 + \ldots + \zeta_n (dx^n)^2 = \sum_{i=1}^{n} \zeta_i (dx^i)^2 \qquad (58)$$

where the indexed $\zeta$ are ±1 while the indexed $x$ are the coordinates of the space. For the space to be flat (i.e. globally not just locally), the condition given by Eq. 58 should apply all over the space and not just at certain points or regions.

An example of flat space is the 3D Euclidean space which can be coordinated by a rectangular Cartesian system whose metric tensor is diagonal with all the diagonal elements being +1. Another example is the 4D Minkowski space-time manifold whose metric is diagonal with elements of ±1. All 1D spaces are Euclidean and hence they cannot be

## 1.4.6 Flat and Curved Spaces

curved intrinsically, so twisted curves are curved only when viewed externally from the embedding space which they reside in, e.g. the 2D space of a surface curve or the 3D space of a space curve. An example of curved space is the 2D surface of a sphere or an ellipsoid.

A necessary and sufficient condition for an $n$D space to be intrinsically flat is that the Riemann-Christoffel curvature tensor (see § 1.4.10) of the space vanishes identically. Hence, cylinders are intrinsically flat, since their Riemann-Christoffel curvature tensor vanishes identically, although they are curved as seen extrinsically from the embedding 3D space. On the other hand, planes are intrinsically and extrinsically flat. In brief, a space is intrinsically flat *iff* the Riemann-Christoffel curvature tensor vanishes identically over the space, and it is extrinsically (as well as intrinsically) flat *iff* the curvature tensor (see § 3.4) vanishes identically over the space. This is because the Riemann-Christoffel curvature tensor characterizes the space curvature from an intrinsic perspective while the curvature tensor characterizes the space curvature from an extrinsic perspective.

Due to the strong connection between the Gaussian curvature (see § 4.5) and the Riemann-Christoffel curvature tensor which implies that each one of these will vanish if the other does (see for example Eq. 92), we see that having an identically vanishing Gaussian curvature is another sufficient and necessary condition for a 2D space to be flat. It should be remarked that the Gaussian curvature in differential geometry is defined for 2D spaces although the concept may be extended to higher dimensionality manifolds in the form of Riemannian curvature. This issue will be discussed further later in the book (see § 4.4).

A curved space may have constant non-vanishing curvature all over the space, or have variable curvature and hence the curvature is position dependent. An example of a space of constant curvature is the surface of a sphere of radius $R$ whose curvature (i.e. Gaussian curvature) is $\frac{1}{R^2}$ at each point of the surface. Torus (Fig. 3), ellipsoid (Fig. 4) and paraboloids (Figs. 7 and 8) are simple examples of surfaces with variable curvature. As we will see, there are various characterizations and quantifications for the curvature; hence in the present context "curvature" may be a generic term. For 2D surfaces, curvature usually refers to the Gaussian curvature (see § 4.5) which is strongly linked to the Riemannian curvature, as indicated above.

Schur theorem related to $n$D spaces ($n > 2$) of constant curvature states that: if the Riemann-Christoffel curvature tensor at each point of a space is a function of the coordinates only, then the curvature is constant all over the space. Schur theorem may also be stated as: the Riemannian curvature is constant over an isotropic region of an $n$D ($n > 2$) Riemannian space. The focus of the book, however, is limited to spaces of lower dimensionality.

The geometry of curved spaces is usually described as the Riemannian geometry. One approach for investigating the Riemannian geometry of a curved manifold is to embed the manifold in a Euclidean space of higher dimensionality and inspect the properties of the manifold from this perspective. This approach is largely followed in the present book where the geometry of curved 2D spaces (twisted surfaces) is investigated by immersing the surfaces in a 3D Euclidean space and examining their properties as viewed from this external enveloping 3D space. Such an external view is necessary for examining the extrinsic geometry of the surface but not its intrinsic geometry. A similar approach is also followed in the investigation of surface and space curves.

A surface with positive/negative Gaussian curvature (see § 4.5) at each point is described as a surface of positive/negative curvature. Ellipsoids (Fig. 4), hyperboloids of two sheets (Fig. 6) and elliptic paraboloids (Fig. 7) are examples of surfaces of positive curvature while hyperboloids of one sheet (Fig. 5) and hyperbolic paraboloids (Fig. 8) are examples of surfaces of negative curvature. A surface with positive/negative Gaussian curvature at each point may also be described as a surface of constant curvature since its sign is constant all over the surface although its magnitude may be variable. In general, a twisted surface possesses coordinate-dependent curvature which varies in magnitude and sign and may take all possible signs (i.e. $< 0$, $0$ and $> 0$) over different points or regions.

The geometric description and quantification of flat spaces are simpler than those of curved spaces, and hence in general the differential geometry of flat spaces is more motivating and less challenging than that of curved spaces. However, as there is a subjective element in this type of statements, it may not apply to everyone.

## 1.4.7 Homogeneous Coordinate Systems

When all the diagonal elements of a diagonal metric tensor of a flat space are $+1$, the coordinate system is described as homogeneous. In this case the line element $ds$ of Eq. 58 becomes:

$$(ds)^2 = dx^i dx^i \tag{59}$$

An example of homogeneous coordinate systems is the rectangular Cartesian system $(x, y, z)$ of a 3D Euclidean space (Fig. 21). A homogeneous coordinate system can be transformed to another homogeneous coordinate system only by linear transformations. Any coordinate system obtained from a homogeneous coordinate system by an orthogonal transformation is also homogeneous. As a consequence of the last statements, infinitely many homogeneous coordinate systems can be constructed in any flat space.

A coordinate system of a flat space can always be homogenized by allowing the coordinates to be imaginary. This is done by redefining the coordinates as:

$$\underline{x}^i = \sqrt{\zeta_i} x^i \tag{60}$$

where the new coordinates $\underline{x}^i$ are imaginary when $\zeta_i = -1$. Consequently, the line element will be given by:

$$(ds)^2 = d\underline{x}^i d\underline{x}^i \tag{61}$$

which is of the same form as Eq. 59. An example of a homogeneous coordinate system with some real and some imaginary coordinates is the coordinate system of a Minkowski 4D space-time related to the mechanics of Lorentz transformations.

## 1.4.8 Geodesic Coordinates

It is always possible to introduce coordinates at particular points in a multi-dimensional manifold so that the Christoffel symbols (see § 1.4.9) vanish at these points. These co-

### 1.4.8 Geodesic Coordinates

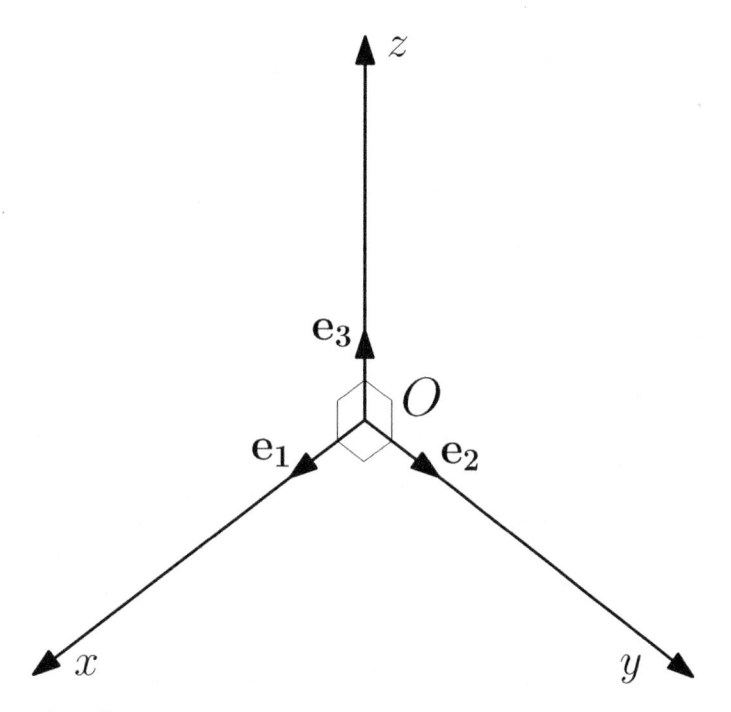

Figure 21: Rectangular Cartesian coordinate system and its basis vectors $\mathbf{e}_1, \mathbf{e}_2$ and $\mathbf{e}_3$ in a 3D space.

ordinates are called geodesic coordinates. Geodesic coordinates are employed as local coordinate systems mainly for the purpose of achieving certain advantages, as will be outlined next. These geodesic systems for these particular points are also described as locally Cartesian coordinates. The allocated points at which the Christoffel symbols are made to vanish in these coordinates are described as the poles. Based on the above and what we will see later in the book, geodesic coordinates of space surfaces are normally taken as the equivalent of the Cartesian coordinates of a flat space. Hence, non-geodesic coordinates on 2D spaces may be compared to general curvilinear coordinates in general $n$D spaces.

The main reason for the use of geodesic coordinates is that the covariant and absolute derivatives (see § 7) in such systems become respectively partial and total derivatives at the poles since the Christoffel symbol terms in the covariant and absolute derivative expressions vanish at these points. Any tensor property can then be easily proved in the geodesic system at the pole and consequently generalized to other systems due to the invariance of the zero tensor under permissible coordinate transformations. If the allocated pole is a general point in the space, the property is then established over the whole space.

In any Riemannian space it is always possible to find a coordinate system for which the coordinates are geodesic at every point of a given analytic curve. Moreover, there is an infinite number of ways by which geodesic coordinates can be defined over a coordinate patch (see § 3.1). We should remark that some authors define geodesic coordinates on a coordinate patch of a surface as a coordinate system whose $u$ and $v$ coordinate curve families are orthogonal, with one of these families ($u$ or $v$) being a family of geodesic curves (refer to § 5.7). Hence, "geodesic coordinates" may appear to have multiple usage

*1.4.9 Christoffel Symbols for Curves and Surfaces* 34

although they are essentially the same. More details about this should be looked for in more advanced textbooks on differential geometry.

## 1.4.9. Christoffel Symbols for Curves and Surfaces

The Christoffel symbols of the first kind for a general $n$D space ($n > 1$) are defined by:

$$[ij, l] = \frac{1}{2} \left( \partial_j g_{il} + \partial_i g_{jl} - \partial_l g_{ij} \right) \qquad (62)$$

where the indexed $g$ is the covariant form of the metric tensor of the given space. We remark that the Latin indices used in the previous equation, as well as the next four equations, do not imply a 3D space and hence the equations and the metric are not specific to such a space as they apply to any $n$D space ($n \geq 2$).

The Christoffel symbols of the second kind, which may also be called affine connections, are obtained by raising the third index of the Christoffel symbols of the first kind, that is:

$$\Gamma^k_{ij} = g^{kl} [ij, l] = \frac{g^{kl}}{2} \left( \partial_j g_{il} + \partial_i g_{jl} - \partial_l g_{ij} \right) \qquad (63)$$

where the indexed $g$ is the metric tensor of the given space in its contravariant and covariant forms with implied summation over $l$. Similarly, the Christoffel symbols of the first kind can be obtained from the Christoffel symbols of the second kind by reversing the above process through lowering the upper index, that is:

$$g_{km} \Gamma^k_{ij} = g_{km} g^{kl} [ij, l] = \delta^l_m [ij, l] = [ij, m] \qquad (64)$$

where $\delta^l_m$ is the mixed form of the Kronecker delta tensor for the space. It is noteworthy that the Christoffel symbols of the first and second kind are symmetric in their paired indices, that is:

$$[ij, k] = [ji, k] \qquad (65)$$
$$\Gamma^k_{ij} = \Gamma^k_{ji} \qquad (66)$$

For 1D spaces, the Christoffel symbols are not defined. The Christoffel symbols of the first kind for a 2D surface are given by:

$$[11, 1] = \frac{\partial_u a_{11}}{2} = \frac{E_u}{2} \qquad (67)$$

$$[11, 2] = \partial_u a_{12} - \frac{\partial_v a_{11}}{2} = F_u - \frac{E_v}{2} \qquad (68)$$

$$[12, 1] = \frac{\partial_v a_{11}}{2} = \frac{E_v}{2} = [21, 1] \qquad (69)$$

$$[12, 2] = \frac{\partial_u a_{22}}{2} = \frac{G_u}{2} = [21, 2] \qquad (70)$$

$$[22, 1] = \partial_v a_{12} - \frac{\partial_u a_{22}}{2} = F_v - \frac{G_u}{2} \qquad (71)$$

### 1.4.9 Christoffel Symbols for Curves and Surfaces

$$[22, 2] = \frac{\partial_v a_{22}}{2} = \frac{G_v}{2} \tag{72}$$

where the indexed $a$ are the elements of the surface covariant metric tensor (refer to 3.3) and $E, F, G$ are the coefficients of the first fundamental form (refer to 3.5). The subscripts $u$ and $v$ which suffix the coefficients stand for partial derivatives of these coefficients with respect to these variables (i.e. $\frac{\partial}{\partial u}$ and $\frac{\partial}{\partial v}$). As indicated before and will be seen in the future, $E = a_{11}$, $F = a_{12} = a_{21}$ and $G = a_{22}$, and hence the equalities in the previous equations are justified. The above relations can be obtained from the definition of the Christoffel symbols of the first kind (Eq. 62) using the coefficients of the metric tensor and first fundamental form of the surface.

In orthogonal coordinate systems, $F = a_{12} = a_{21} = 0$ identically and hence the above formulae will be simplified accordingly by dropping any term involving the derivatives of these coefficients. We note that "orthogonal coordinate system" here and in similar contexts means a system whose coordinate curves are orthogonal everywhere. Therefore, the above condition is fully justified since $\mathbf{E}_1$ and $\mathbf{E}_2$ basis vectors, which are the tangents to the $u$ and $v$ coordinate curves, will be orthogonal in such a system and hence the dot product, seen in Eq. 235, will vanish accordingly.

The Christoffel symbols of the second kind for a 2D surface are given by:

$$\Gamma^1_{11} = \frac{a_{22}\partial_u a_{11} - 2a_{12}\partial_u a_{12} + a_{12}\partial_v a_{11}}{2a} = \frac{GE_u - 2FF_u + FE_v}{2a} \tag{73}$$

$$\Gamma^2_{11} = \frac{2a_{11}\partial_u a_{12} - a_{11}\partial_v a_{11} - a_{12}\partial_u a_{11}}{2a} = \frac{2EF_u - EE_v - FE_u}{2a} \tag{74}$$

$$\Gamma^1_{12} = \frac{a_{22}\partial_v a_{11} - a_{12}\partial_u a_{22}}{2a} = \frac{GE_v - FG_u}{2a} = \Gamma^1_{21} \tag{75}$$

$$\Gamma^2_{12} = \frac{a_{11}\partial_u a_{22} - a_{12}\partial_v a_{11}}{2a} = \frac{EG_u - FE_v}{2a} = \Gamma^2_{21} \tag{76}$$

$$\Gamma^1_{22} = \frac{2a_{22}\partial_v a_{12} - a_{22}\partial_u a_{22} - a_{12}\partial_v a_{22}}{2a} = \frac{2GF_v - GG_u - FG_v}{2a} \tag{77}$$

$$\Gamma^2_{22} = \frac{a_{11}\partial_v a_{22} - 2a_{12}\partial_v a_{12} + a_{12}\partial_u a_{22}}{2a} = \frac{EG_v - 2FF_v + FG_u}{2a} \tag{78}$$

where $a$ $(= a_{11}a_{22} - a_{12}a_{21} = EG - F^2)$ is the determinant of the surface covariant metric tensor, and the other symbols are as explained above. These relations can be obtained from the definition of the Christoffel symbols of the second kind (Eq. 63) using the coefficients of the metric tensor and first fundamental form of the surface. The above formulae will also be simplified in orthogonal coordinate systems, where $F = a_{12} = a_{21} = 0$ identically, by dropping the vanishing terms that involve these coefficients or their derivatives.

The Christoffel symbols of the second kind for a 2D surface may also be given by:

$$\Gamma^1_{11} = -\frac{(\mathbf{E}_2 \times \partial_1 \mathbf{E}_1) \cdot \mathbf{n}}{\sqrt{a}} \tag{79}$$

$$\Gamma^2_{11} = +\frac{(\mathbf{E}_1 \times \partial_1 \mathbf{E}_1) \cdot \mathbf{n}}{\sqrt{a}} \tag{80}$$

### 1.4.10 Riemann-Christoffel Curvature Tensor

$$\Gamma^1_{12} = -\frac{(\mathbf{E}_2 \times \partial_2 \mathbf{E}_1) \cdot \mathbf{n}}{\sqrt{a}} = \Gamma^1_{21} \tag{81}$$

$$\Gamma^2_{12} = +\frac{(\mathbf{E}_1 \times \partial_2 \mathbf{E}_1) \cdot \mathbf{n}}{\sqrt{a}} = \Gamma^2_{21} \tag{82}$$

$$\Gamma^1_{22} = -\frac{(\mathbf{E}_2 \times \partial_2 \mathbf{E}_2) \cdot \mathbf{n}}{\sqrt{a}} \tag{83}$$

$$\Gamma^2_{22} = +\frac{(\mathbf{E}_1 \times \partial_2 \mathbf{E}_2) \cdot \mathbf{n}}{\sqrt{a}} \tag{84}$$

where the indexed $\mathbf{E}$ are the surface covariant basis vectors (see § 1.4.5 and 3.2), $\mathbf{n}$ is the unit normal vector to the surface and $a$ is the determinant of the surface covariant metric tensor as defined above. It is worth noting that: $\sqrt{a} = (\mathbf{E}_1 \times \mathbf{E}_2) \cdot \mathbf{n}$, as we will see later in the book.

Since the Christoffel symbols of both kinds are dependent on the metric only, as can be seen from the above equations (Eqs. 67-72 and Eqs. 73-78) as well as from the definitions of these symbols (see Eqs. 62 and 63), they represent intrinsic properties of the surface geometry and hence they are part of its intrinsic geometry. The involvement of extrinsic parameters in the definitions given by Eqs. 79-84 does not affect their intrinsic qualification, as explained earlier.

The Christoffel symbols of the first kind are linked to the surface covariant basis vectors by the following relation:

$$[\alpha\beta, \gamma] = \frac{\partial \mathbf{E}_\alpha}{\partial u^\beta} \cdot \mathbf{E}_\gamma \qquad (\alpha, \beta, \gamma = 1, 2) \tag{85}$$

This relation may also be written as:

$$[\alpha\beta, \gamma] = \mathbf{r}_{\alpha\beta} \cdot \mathbf{r}_\gamma \qquad (\alpha, \beta, \gamma = 1, 2) \tag{86}$$

where the subscripts represent partial derivatives with respect to the variables represented by these coordinate indices. The last equation may provide an easier form to remember these formulae.

On applying the index raising operator to Eq. 85, we obtain a similar expression for the Christoffel symbols of the second kind, that is:

$$\Gamma^\gamma_{\alpha\beta} = \frac{\partial \mathbf{E}_\alpha}{\partial u^\beta} \cdot \mathbf{E}^\gamma \qquad (\alpha, \beta, \gamma = 1, 2) \tag{87}$$

where $\mathbf{E}^\gamma$ is the contravariant form of the surface basis vectors (see § 3.2).

### 1.4.10 Riemann-Christoffel Curvature Tensor

The Riemann-Christoffel curvature tensor is an absolute rank-4 tensor that characterizes important properties of spaces, including 2D surfaces, and hence it plays an important role in differential geometry. The tensor is used, for instance, to test for the space flatness (see § 1.4.6). There are two kinds of Riemann-Christoffel curvature tensor: first and second.

## 1.4.10 Riemann-Christoffel Curvature Tensor

The Riemann-Christoffel curvature tensor of the first kind is a type $(0,4)$ tensor while the Riemann-Christoffel curvature tensor of the second kind is a type $(1,3)$ tensor. Shifting from one kind to the other is achieved by using the index shifting operator. The first and second kinds of the Riemann-Christoffel curvature tensor are given respectively by:

$$R_{ijkl} = \partial_k [jl,i] - \partial_l [jk,i] + [il,r]\Gamma^r_{jk} - [ik,r]\Gamma^r_{jl} \tag{88}$$

$$R^i_{jkl} = \partial_k \Gamma^i_{jl} - \partial_l \Gamma^i_{jk} + \Gamma^r_{jl}\Gamma^i_{rk} - \Gamma^r_{jk}\Gamma^i_{rl} \tag{89}$$

where the indexed square brackets and $\Gamma$ are the Christoffel symbols of the first and second kind of the given space, as defined in § 1.4.9.

We remark that in the above two equations, as well as in the following equations in the present and the next subsection, the Latin indices do not necessarily range over $1,2,3$ as these equations are valid for general $n$D manifolds ($n \geq 2$) including surfaces ($n = 2$) and spaces of higher dimensionality ($n > 3$). Another remark is that the Riemann-Christoffel curvature tensor of the first kind is anti-symmetric in its first two indices and in its last two indices and block symmetric with respect to these two sets of indices, that is:

$$R_{ijkl} = -R_{jikl} \qquad R_{ijkl} = -R_{ijlk} \qquad R_{ijkl} = R_{klij} \tag{90}$$

From Eqs. 88 and 89, it can be seen that the Riemann-Christoffel curvature tensor depends exclusively on the Christoffel symbols of the first and second kind which are both dependent on the metric tensor (or the first fundamental form, see § 3.5) and its partial derivatives only. Hence, the Riemann-Christoffel curvature, as represented by the Riemann-Christoffel curvature tensor, is an intrinsic property of the manifold. Since the Riemann-Christoffel curvature tensor depends on the metric which, in general curvilinear coordinates, is a function of position, the Riemann-Christoffel curvature tensor follows this dependency on position.

The Riemann-Christoffel curvature tensor vanishes identically *iff* the space is globally flat from intrinsic view; otherwise the space is curved (see § 1.4.6). Hence, the Riemann-Christoffel curvature tensor vanishes identically over 1D manifolds as represented by surface and space curves. As we will see in § 2.3, surface and space curves are intrinsically Euclidean. Similarly, the Riemann-Christoffel curvature tensor vanishes identically over plane surfaces. More generally, a surface is isometric to the Euclidean plane *iff* the Riemann-Christoffel curvature tensor is zero at each point on the surface since it is intrinsically flat. The last statement also applies if the Gaussian curvature of the surface vanishes identically due to the link between the Riemann-Christoffel curvature tensor and the Gaussian curvature (refer to § 4.5 and see Eq. 92).

The 2D Riemann-Christoffel curvature tensor has only one degree of freedom and hence it possesses a single independent non-vanishing component which is represented by $R_{1212}$. Hence, for a 2D Riemannian space we have:

$$R_{1212} = R_{2121} = -R_{1221} = -R_{2112} \tag{91}$$

while all the other components of the tensor are identically zero. The signs in the equalities of Eq. 91 are justified by the aforementioned anti-symmetric relations of the Riemann-Christoffel tensor in its indices (see Eq. 90). The vanishing of the other components (e.g. $R_{1111}$ and $R_{1121}$) can also be explained by the anti-symmetric relations since these components contain two identical anti-symmetric indices.

Based on the above facts, the Riemann-Christoffel curvature tensor can be expressed in tensor notation by:

$$R_{\alpha\beta\gamma\delta} = R_{1212}\epsilon_{\alpha\beta}\epsilon_{\gamma\delta} = \frac{R_{1212}}{a}\underline{\epsilon}_{\alpha\beta}\underline{\epsilon}_{\gamma\delta} = K\underline{\epsilon}_{\alpha\beta}\underline{\epsilon}_{\gamma\delta} \qquad (92)$$

where the indexed $\epsilon$ are the relative permutation tensors, the indexed $\underline{\epsilon}$ are the absolute permutation tensors, and $K$ is the Gaussian curvature (see § 4.5) whose expression is obtained from Eq. 356. The non-vanishing component of the 2D Riemann-Christoffel curvature tensor, $R_{1212}$, may be given in expanded form by:

$$R_{1212} = \frac{1}{2}\left(2\partial_{12}a_{12} - \partial_{22}a_{11} - \partial_{11}a_{22}\right) + a_{\alpha\beta}\left(\Gamma^{\alpha}_{12}\Gamma^{\beta}_{12} - \Gamma^{\alpha}_{11}\Gamma^{\beta}_{22}\right) \qquad (93)$$

where $\partial_{\alpha\beta} \equiv \frac{\partial^2}{\partial u^\alpha \partial u^\beta}$ with $\alpha, \beta = 1, 2$, the indexed $a$ are the coefficients of the surface covariant metric tensor, and the Christoffel symbols are based on the surface metric.

### 1.4.11 Ricci Curvature Tensor and Scalar

The Ricci curvature tensor of the first kind, which is an absolute rank-2 covariant symmetric tensor, is obtained by contracting the contravariant index with the last covariant index of the Riemann-Christoffel curvature tensor of the second kind, that is:

$$R_{ij} = R^{a}_{ija} = \partial_j \Gamma^{a}_{ia} - \partial_a \Gamma^{a}_{ij} + \Gamma^{a}_{bj}\Gamma^{b}_{ia} - \Gamma^{a}_{ba}\Gamma^{b}_{ij} \qquad (94)$$

The Ricci tensor of the second kind is obtained by raising the first index of the Ricci tensor of the first kind using the index raising operator.

For 2D spaces, the Ricci curvature tensor of the first kind is related to the Riemann-Christoffel curvature tensor by the following relations:

$$\frac{R_{11}}{a_{11}} = \frac{R_{12}}{a_{12}} = \frac{R_{21}}{a_{21}} = \frac{R_{22}}{a_{22}} = -\frac{R_{1212}}{a} \qquad (95)$$

where the indexed $a$ are the elements of the 2D covariant metric tensor (see § 3.3) and the bare $a$ is its determinant. As the 2D Riemann-Christoffel curvature tensor has only one independent non-vanishing component, the last equation provides a full link between the Ricci curvature tensor and the Riemann-Christoffel curvature tensor. Since $K = \frac{R_{1212}}{a}$ (see Eq. 356), the above relations also link the Ricci tensor to the Gaussian curvature.

The Ricci scalar $\mathcal{R}$, which is also called the curvature scalar and the curvature invariant, is the result of contracting the indices of the Ricci curvature tensor of the second kind, that is:

$$\mathcal{R} = R^{i}_{i} \qquad (96)$$

Hence, it is a rank-0 tensor. As seen from the above relations, the Ricci curvature tensor and curvature scalar are part of the intrinsic geometry of the manifold.

## 1.5 Exercises

1.1 Give a brief definition of differential geometry indicating the other disciplines of mathematics to which differential geometry is intimately linked.

1.2 A surface embedded in a 3D space can be regarded as a 2D and as a 3D object at the same time. Discuss this briefly. From the same perspective, discuss also the state of a curve embedded in a surface which in its turn is embedded in a 3D space.

1.3 What are the following symbols: $[\alpha\beta, \gamma]$, $[ij, k]$, $\Gamma^{\gamma}_{\alpha\beta}$ and $\Gamma^{k}_{ij}$? What is the difference between those with Greek indices and those with Latin indices?

1.4 What is the relation between the coefficients of the surface covariant metric tensor and the surface covariant curvature tensor on one hand and the coefficients of the first and second fundamental forms on the other? What are the symbols representing all these coefficients?

1.5 What is the difference between the local and global properties of a manifold? Give an example for each. What are the colloquial terms used to label these two categories?

1.6 What is the meaning of "intrinsic" and "extrinsic" properties of a manifold? Give an example for each.

1.7 Explain the concept of "2D inhabitant" and how it is used to classify the properties of a space surface.

1.8 Find the equation of a plane passing through the points: $(1, 2, 0)$, $(0, -3, 1.5)$ and $(1, 0, -1)$. What is the normal unit vector to this plane?

1.9 Is the normal unit vector of a plane surface unique?

1.10 Define briefly each one of the following terms: surface of revolution, meridians and parallels.

1.11 Prove that the meridians and parallels of a surface of revolution are mutually perpendicular at their points of intersection.

1.12 State the parametric equations of the following geometric shapes: torus, hyperboloid of one sheet, and hyperbolic paraboloid.

1.13 Write down the parametric equations of a circle in the $uv$ plane centered at point $(a, b)$ with radius $r = c$ where $a, b, c$ are real constants and $c > 0$.

1.14 Find the parametric equations of an ellipse in the $xy$ plane centered at the origin of coordinates with $A = 5$ and $B = 3$ where $A, B$ are the semi-major and semi-minor axes.

1.15 A surface of revolution may be represented locally in a 3D space by the following form: $\mathbf{r}(u, v) = (u \cos v, u \sin v, f(u, v))$ where $f$ is a continuous function. Determine the equations representing the parallels and meridians of this surface.

1.16 Find the parametric equations of a curve formed by the intersection of the surfaces represented by: $\mathbf{r}_1(u, v) = (u, u^2, v)$ and $\mathbf{r}_2(u, v) = (u, v, u^2)$ where $-\infty < u, v < +\infty$.

1.17 Write down the general form of the parametric equations of each of the following surfaces: hyperboloid of two sheets, parabolic cylinder, catenoid, monkey saddle and pseudo-sphere.

1.18 Sketch the following (using a 3D computer graphic package if available): (a) a straight line passing through the point $(11, -5, 6.3)$ and parallel to the vector $(-3, -1.8, 6.5)$

## 1.5 Exercises

(b) a plane passing through the point $(6, -8.2, -7)$ with a normal vector $(3, -1.6, -2.5)$.

1.19 A surface is parameterized by: $\mathbf{r}(u,v) = (a\sinh u \cos v, b \sinh u \sin v, c \cosh u)$. What is the name of this surface? What is the condition for this surface to be a surface of revolution around the third spatial axis?

1.20 Find the equation of the straight line passing through the point $(-6, 3.1, 8.4)$ and the point $(1, 0, -3)$.

1.21 Classify the following as curves or surfaces: ellipsoid, elliptic paraboloid, catenary, helicoid, enneper and tractrix.

1.22 Make a simple sketch for each one of the geometric shapes in the previous question. Use a computer graphic package if convenient.

1.23 Prove that $\mathbf{v} \times \frac{d\mathbf{v}}{dt} = \mathbf{0}$ iff the direction of the vector $\mathbf{v}(t)$ is constant.

1.24 Define "Euler characteristic" stating the equation that links it to the number of vertices, faces and edges of a polyhedron.

1.25 Explain how the Euler characteristic is defined for non-polyhedral compact orientable surfaces such as ellipsoids.

1.26 What is the topological meaning of "genus of a surface"?

1.27 Give examples for surfaces of genus 0, 1, 2 and 3 from common geometric shapes other than those given in the text.

1.28 By using the Euler formula, calculate the Euler characteristic $\chi$ of the following surfaces: (a) parallelepiped (b) dodecahedron (c) icosahedron. Show your work in detail.

1.29 By using polygonal decomposition, calculate the Euler characteristic $\chi$ of the following surfaces: (a) sphere (b) ellipsoid (c) torus. Show your work in detail with simple sketches to demonstrate the polygonal decomposition in each case.

1.30 What is the genus of the surfaces in the previous question?

1.31 What is the Cartesian form of the equation of a sphere centered at point $(a, b, c)$ with radius $r = d$ where $a, b, c, d$ are real constants and $d > 0$?

1.32 Explain briefly the meaning of the following terms: bicontinuous function, surface of class $C^n$, and sufficiently smooth curve.

1.33 Explain in detail, using equations and simple sketches, the concept of "deleted neighborhood" in 1D and 2D flat and curved spaces as seen from the ambient space.

1.34 How can we extend the concept of "deleted neighborhood" to spaces of dimensionality higher than 2?

1.35 Find the equation of a cone generated by rotating the line $z = -2x$ around the $z$-axis.

1.36 Derive the parametric equations of a helix rotating around the $z$-axis, passing through the point $(3, 0, 0)$ and climbing (or descending) 5.3 units in the $z$-direction as it makes a $4\pi$ turn around the $z$-axis.

1.37 Define "positive definite" in words and by stating the mathematical conditions for a quadratic expression to be positive definite.

1.38 Describe orthogonal coordinate transformations and how they are characterized by their Jacobian.

1.39 State the difference between positive and negative orthogonal transformations.

1.40 Find a set of parametric equations representing a cylinder generated by rotating a straight line parallel to the $z$-axis and passing through the point (2.5,0,0) around the

## 1.5 Exercises

$z$-axis.

1.41 What is the unit normal vector to the surface of a sphere, centered on the origin of coordinates with radius $r$, at a point on its surface with coordinates $(x_P, y_P, z_P)$? Consider the possibility of having more than one normal vector at that point.

1.42 What "coordinate curves" means? What are the other names given to these curves?

1.43 Define regular representation of a class $C^n$ surface patch in a 3D Euclidean space stating its mathematical conditions.

1.44 Using a parametric representation of the elliptic paraboloid, show that it is a regular surface.

1.45 What are the reasons for having a singular point on a space surface?

1.46 A sphere centered at the origin of coordinates can be represented parametrically by: $\mathbf{r}(\theta, \phi) = a(\sin\theta\cos\phi, \sin\theta\sin\phi, \cos\theta)$ where $a > 0$ is a constant, $0 \leq \theta \leq \pi$ and $0 \leq \phi < 2\pi$. At what points, if any, this representation is not regular?

1.47 How a mathematical correspondence can be established between points on two different curves and two different surfaces?

1.48 State the mathematical conditions which are satisfied by the intrinsic distance between two points on a smooth connected surface.

1.49 Is it guaranteed that an arc of minimum length between two specific points on a surface does exist and it is unique?

1.50 Prove the three properties of intrinsic distance, i.e. symmetry, triangle inequality, and positive definiteness.

1.51 Show that the intrinsic distance $d$ between two points is invariant under a local isometric mapping $f$, i.e. $d(f(P_1), f(P_2)) = d(P_1, P_2)$.

1.52 Find the intrinsic distance on the unit sphere centered at the origin of coordinates between the point $(1, 0, 0)$ and the point $(\frac{1}{\sqrt{3}}, \frac{1}{\sqrt{3}}, \frac{1}{\sqrt{3}})$.

1.53 What is the intrinsic distance between the two points of the last question in a 3D Euclidean space that encloses the sphere?

1.54 Define the basis vectors and state their roles.

1.55 Describe in detail the two main sets of space basis vectors in differential geometry related to curves and surfaces. Are there any other sets of basis vectors?

1.56 Are the basis vectors necessarily of unit length and/or mutually orthogonal? If not, give examples of basis vectors which are not of unit length and/or mutually orthogonal.

1.57 Define, in mathematical terms, flat and curved spaces giving examples for each.

1.58 State a sufficient and necessary condition for an $n$D space to be flat.

1.59 Is it necessary that an $n$D curved space possesses universally constant curvature? If not, give an example of a space with variable curvature in sign and magnitude.

1.60 What is the locus of the points (if any) which are shared between the $xy$ plane and the following surfaces: (a) a sphere centered at $(0, 0, 5)$ with radius $r = 6$ (b) a sphere centered at $(1, 1, 1)$ with radius $r = 1.5$ (c) a plane passing through the point $(5, -9.6, 0)$ with a unit normal vector $(0, 0, -1)$?

1.61 Describe the commonly used approach for investigating the Riemannian geometry of curved manifolds.

1.62 Give a brief definition of homogeneous coordinate systems giving a common example

## 1.5 Exercises

of such systems.

1.63 What is the relation between the Christoffel symbols of the first kind and the Christoffel symbols of the second kind?

1.64 Write the mathematical expressions for the symbols $[12, 1]$ and $\Gamma^1_{22}$ of a surface in terms of the coefficients of the surface metric tensor.

1.65 Using the definition of the Christoffel symbols of the first kind and the rules of tensors, derive Eqs. 68 and 71.

1.66 Using the definition of the Christoffel symbols of the second kind and the rules of tensors, derive Eqs. 76 and 78.

1.67 State the mathematical relations correlating the Christoffel symbols of the first and second kind to the surface basis vectors and their derivatives.

1.68 What is the relation between the Riemann-Christoffel curvature tensor and the Gaussian curvature of a surface?

1.69 How many independent non-vanishing components the 2D Riemann-Christoffel curvature tensor possesses?

1.70 What is the significance of having an identically vanishing Riemann-Christoffel curvature tensor on a 2D surface?

1.71 Show that the Riemann-Christoffel curvature tensor is anti-symmetric in its first two indices and in its last two indices and block symmetric with respect to these two sets of indices.

1.72 Prove Eqs. 85 and 87.

1.73 What is the rank of the Ricci curvature tensor?

1.74 State the mathematical relation that links the Ricci curvature tensor of the first kind to the Christoffel symbols of the second kind and their partial derivatives.

1.75 What is the relation between the elements of the Ricci curvature tensor of the first kind and the Gaussian curvature?

1.76 How do you obtain the Ricci scalar from the Riemann-Christoffel curvature tensor of the first kind? Explain your answer step by step.

# Chapter 2
# Curves in Space

In this chapter, we investigate curves residing in a higher dimensionality space and how they are characterized. We should first remark that "space" in this title is general and hence it includes surface since it is a 2D space, as explained in § 1.2.

## 2.1 General Background about Curves

In simple terms, a space curve is a set of connected points[7] in the embedding space such that any totally connected subset of it can be twisted into a straight line segment without affecting the neighborhood of any point. More technically, a curve is defined as a differentiable parameterized mapping between an interval of the real line and a connected subset of the embedding space, that is $C(t) : I \to \mathbb{R}^n$ where $C$ represents a space curve defined on the interval $I \subseteq \mathbb{R}$ and parameterized by the variable $t \in I$. Hence, different parameterizations of the same "geometric curve" will lead to different "mapping curves". The image of the mapping in the embedding space is known as the trace of the curve; hence different mapping curves can share the same trace. The curve may also be defined as a topological image of a real interval and may be linked to the concept of Jordan arc. We note that Jordan arcs or Jordan curves may be defined as injective mappings with no self intersection.

Space curves can be defined symbolically in different ways; the most common of these is parametrically where the three space coordinates of the curve points are given as functions of a real valued parameter, e.g. $x^i = x^i(t)$ where $t \in \mathbb{R}$ is the curve parameter and $i = 1, 2, 3$. The parameter $t$ may represent time or arc length or even an arbitrarily defined quantity. Similarly, surface curves are defined parametrically where the two surface coordinates are given as functions of a real valued parameter, e.g. $u^\alpha = u^\alpha(t)$ with $\alpha = 1, 2$. The surface coordinates can then be mapped, through another mapping relation, onto the spatial representation of the surface in the enveloping space. Parameterized curves are oriented objects as they can be traversed in one direction or the other depending on the sense of increase of their parameter.

There are two main types of curve parameterization: parameterization by arc length and parameterization by something else such as time. The condition for a space curve $C(t) : I \to \mathbb{R}^3$, where $t \in I$ is the curve parameter and $I \subseteq \mathbb{R}$ is an interval over which the curve is defined, to be parameterized by arc length is that: for all $t$ we have $|\frac{d\mathbf{r}}{dt}| = 1$ where $\mathbf{r}(t)$ is the position vector representing the curve in the ambient space. As a consequence

---

[7] The points are usually assumed to be totally connected so that any point on the curve can be reached from any other point by passing through other curve points or at least they are piecewise connected. We also consider mostly open curves with simple connectivity and hence the curve does not intersect itself.

## 2.1 General Background about Curves

of this, parameterization by arc length is equivalent to traversing the curve with a constant unity speed. Hence, using a parameterization by something other than arc length may be considered as traversing the curve with varying or non-unity speed. The advantage of parameterization by arc length is that it confines the attention on the geometry of the curve rather than other factors, which are usually irrelevant to the geometric investigation, such as the temporal rate of traversing the curve. Furthermore, it usually results in a more simple mathematical formulation, as we will see later in the book.

The parameter symbol which is used normally for parameterization by arc length is $s$, while $t$ is used to represent a general parameter which could be arc length or something else. This notation is followed in the present book. For curves parameterized by arc length, the length $L$ of a segment between two points on the curve corresponding to $s_1$ and $s_2$ is given by the simple formula:

$$L = \left| \int_{s_1}^{s_2} dt \right| = |s_2 - s_1| \tag{97}$$

Parameterization by arc length $s$ may be called natural parameterization of the curve and hence $s$ is called natural parameter.

Natural parameterization is not unique; however any other natural parameter $\check{s}$ is related to a given natural parameter $s$ by the following relation:

$$\check{s} = \pm s + c \tag{98}$$

where $c$ is a real constant and hence the above-stated condition $|\frac{d\mathbf{r}}{dt}| = 1$ remains valid.[8] This may be stated in a different way by saying that natural parameterization with arc length $s$ is unique apart from the possibility of having a different sense of orientation and an additive constant to $s$.

Natural parameterization may also be used for parameterization by a parameter which is proportional to $s$ and hence the transformation relation between two natural parameters becomes:

$$\check{s} = \pm ms + c \tag{99}$$

where $m$ is another real constant. The two parameterizations then differ, apart from the sense of orientation and the constant shift, by the length scale which can be chosen arbitrarily. Consequently, natural parameterization will be equivalent to traversing the curve with a constant speed not necessarily of unity magnitude. This may be based on the vision of extending the aforementioned benefits of natural parameterization by scaling, i.e. by choosing a different length scale a natural parameterization will be obtained spontaneously. In this book, natural parameterization is restricted to parameterization with arc length $s$.

In a general $n$D space, the tangent vector to a space curve, represented parametrically by the spatial representation $\mathbf{r}(t)$ where $t$ is a general parameter, is given by $\frac{d\mathbf{r}}{dt}$. A vector tangent to a space curve at $P$ is a non-trivial scalar multiple of $\frac{d\mathbf{r}}{dt}$ and hence it can differ in magnitude and direction from $\frac{d\mathbf{r}}{dt}$ as it can be parallel or anti-parallel to $\frac{d\mathbf{r}}{dt}$.

---

[8] In this formula, $t$ is a generic symbol and hence it stands for $s$.

## 2.1 General Background about Curves

The tangent vector to a surface curve, represented parametrically by: $C(u(t), v(t))$ where $u$ and $v$ are the surface coordinates and $t$ is a general parameter, is given by:

$$\frac{d\mathbf{r}}{dt} = \frac{\partial \mathbf{r}}{\partial u}\frac{du}{dt} + \frac{\partial \mathbf{r}}{\partial v}\frac{dv}{dt} \tag{100}$$

where $\mathbf{r}(u(t), v(t))$ is the spatial representation of $C$ and where all these quantities are defined and evaluated at a particular point on the curve. The last equation can be cast compactly, using tensor notation, as:

$$\frac{dx^i}{dt} = \frac{\partial x^i}{\partial u^\alpha}\frac{du^\alpha}{dt} = x^i_\alpha \frac{du^\alpha}{dt} \tag{101}$$

where $i = 1, 2, 3$, $\alpha = 1, 2$ and $(u^1, u^2) \equiv (u, v)$.

A space curve $C(t): I \to \mathbb{R}^3$, where $I \subseteq R$ and $t \in I$ is a general parameter, is "regular at point $t_0$" *iff* $\dot{\mathbf{r}}(t_0)$ exists and $\dot{\mathbf{r}}(t_0) \neq \mathbf{0}$ where $\mathbf{r}(t)$ is the spatial representation of $C$ and the overdot stands for differentiation with respect to the general parameter $t$. The curve is "regular" *iff* it is regular at each interior point in $I$. On a regular parameterized curve there is a neighborhood to each point in its domain in which the curve is injective. On transforming a surface $S$ by a differentiable regular mapping $f$ of class $C^n$ to a surface $\bar{S}$, a regular curve $C$ of class $C^n$ on $S$ will be mapped on a regular curve $\bar{C}$ of class $C^n$ on $\bar{S}$ by the same functional mapping relation, that is $\bar{\mathbf{r}}(t) = f(\mathbf{r}(t))$ where the barred and unbarred $\mathbf{r}(t)$ are the spatial parametric representations of the two curves on the barred and unbarred surfaces.

The tangent line to a sufficiently smooth curve at one of its regular points $P$ is a straight line passing through $P$ but not through any point in a deleted neighborhood of $P$. More technically, the tangent line to a curve $C$ at a given regular point $P$ is a straight line passing through $P$ and having the same orientation as the tangent vector $\frac{d\mathbf{r}}{dt}$ of $C$ at $P$. The tangent line to a curve at a given point $P$ on the curve may also be defined geometrically as the limit of a secant line passing through $P$ and another neighboring point on the curve as the other point converges, while staying on the curve, to the tangent point. These different definitions are equivalent as they represent the same entity. It is noteworthy that the tangent line of a smooth curve, where such a tangent does exit, is unique. We also note that the geometric definition may be useful in some cases where the analytical definition does not apply.

A non-trivial vector $\mathbf{v}$ is said to be tangent to a regular surface $S$ (see § 1.4.3) at a given point $P$ on $S$ if there is a regular curve $C$ on $S$ passing through $P$ such that $\mathbf{v} = \frac{d\mathbf{r}(t)}{dt}$ where $\mathbf{r}(t)$ is the spatial representation of $C$ and $\frac{d\mathbf{r}(t)}{dt}$ is evaluated at $P$ (also see 3.1 for further details). In fact, any vector $\mathbf{v} = c\frac{d\mathbf{r}(t)}{dt}$, where $c \neq 0$ is a real number, is a tangent although it may not be *the* tangent.

A periodic curve $C$ is a curve that can be represented parametrically by a continuous function of the form $\mathbf{r}(t + T) = \mathbf{r}(t)$ where $\mathbf{r}$ is the spatial representation of $C$, $t$ is a real general parameter and $T$ is a real constant called the function period. Circles and ellipses (Fig. 22) are prominent examples of periodic curves where they can be represented parametrically by:

$$\mathbf{r}(t) = (a\cos t, a\sin t) \quad \text{(circle)} \quad (102)$$
$$\mathbf{r}(t) = (a\cos t, b\sin t) \quad \text{(ellipse)} \quad (103)$$

where $a$ and $b$ are real positive constants and $t \in \mathbb{R}$. Due to the periodicity of the trigonometric functions, these equations satisfy the condition: $\mathbf{r}(t + 2\pi) = \mathbf{r}(t)$. Hence, circles and ellipses are periodic curves with a period of $2\pi$.

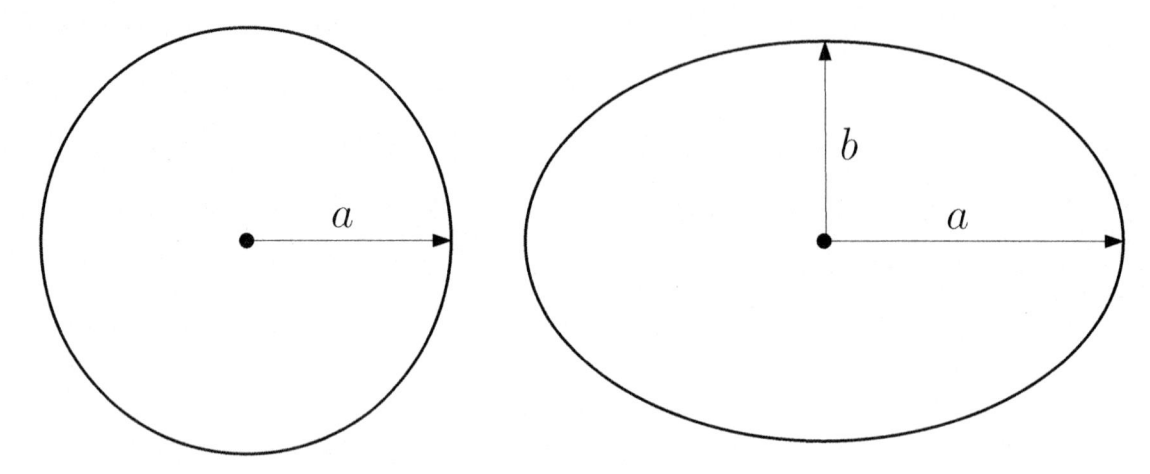

Figure 22: Circle (left frame) and ellipse (right frame) and their main parameters.

A closed curve is a continuous periodic curve defined over a minimum of one period. We note that periodicity is not a necessary requirement for the definition of closed curves as the curves can be defined over a single period without being considered as such or by functions of non-periodic nature. Closed curves may be regarded as topological images of circles.

A curve is described as a plane curve if it can be embedded entirely in a plane with no distortion. Orthogonal trajectories of a given family of curves is a family of curves that intersect the given family perpendicularly at their intersection points. Any curve can be mapped isometrically to a straight line segment where both are naturally parameterized by arc length. From the last statement plus the fact that isometric transformation is an equivalence relation (see § 6.5), it can be concluded that any two space and surface curves can be connected by an isometric relation.

## 2.2 Mathematical Description of Curves

Let have a space curve of class $C^2$ in a 3D Riemannian manifold with a given metric $g_{ij}$ ($i, j = 1, 2, 3$). The curve is parameterized naturally by $s$ representing arc length. As stated earlier, we choose to parameterize the curve by $s$ to have simpler formulae although, for the sake of completeness and generality, other formulae based on a more general parameterization will also be given. The curve can therefore be represented by:

$$x^i = x^i(s) \quad (i = 1, 2, 3) \quad (104)$$

## 2.2 Mathematical Description of Curves

where the indexed $x$ represent the space coordinates. This is equivalent to:

$$\mathbf{r}(s) = x^i(s)\mathbf{E}_i \tag{105}$$

where $\mathbf{r}$ is the spatial representation of the space curve and $\mathbf{E}_i$ are the space basis vectors.

Three mutually perpendicular vectors each of unit length can be defined at each regular point of the above-described space curve: tangent $\mathbf{T}$, normal $\mathbf{N}$ and binormal $\mathbf{B}$ (see Fig. 23). As well as characterizing the curve, these vectors can serve as a moving coordinate system for the embedding space as indicated earlier. For simplicity, clarity and potential lack of familiarity with tensor differentiation (see § 7) at this stage, the following is mostly based on assuming a Euclidean space coordinated by a rectangular Cartesian system although supplementary remarks related to more general space and coordinates are added when necessary. We also use a mix of tensor and symbolic notations as each has certain advantages and to familiarize the reader with both notations since different authors use different notations.

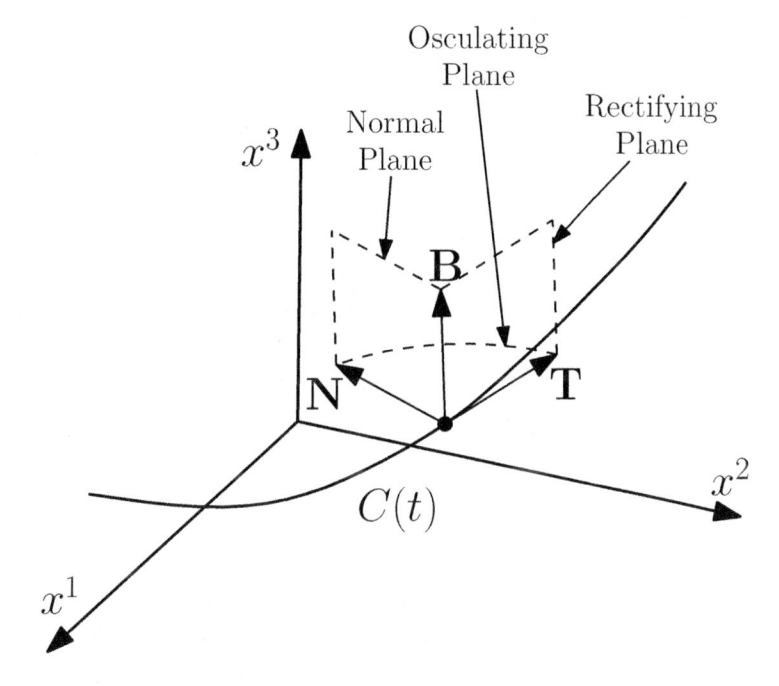

Figure 23: The vectors $\mathbf{T}, \mathbf{N}, \mathbf{B}$ and their associated planes of a curve $C$ embedded in a Euclidean space with a rectangular Cartesian coordinate system (see also Fig. 24).

The unit vector tangent to the curve at a given regular point $P$ on the curve is given by:[9]

$$[\mathbf{T}]^i = T^i = \frac{dx^i}{ds} \tag{106}$$

---

[9] Since we employ Cartesian coordinates in a flat space, ordinary derivatives (i.e. $\frac{d}{ds}$ and $\frac{d}{dt}$) are used in this and the following formulae. For general curvilinear coordinates, these ordinary derivatives should be replaced by absolute derivatives (i.e. $\frac{\delta}{\delta s}$ and $\frac{\delta}{\delta t}$) along the curves.

## 2.2 Mathematical Description of Curves

For a $t$-parameterized curve, where $t$ is not necessarily the arc length, the tangent vector is given by:

$$\mathbf{T} = \frac{\dot{\mathbf{r}}(t)}{|\dot{\mathbf{r}}(t)|} \qquad (107)$$

where the overdot represents differentiation with respect to $t$.

The unit vector normal to the tangent $T^i$, and hence to the curve, at the point $P$ is given by:

$$[\mathbf{N}]^i = N^i = \frac{\frac{dT^i}{ds}}{\left|\frac{dT^i}{ds}\right|} = \frac{1}{\kappa}\frac{dT^i}{ds} \qquad (108)$$

where $\kappa$ is a scalar called the "curvature" of the curve at the point $P$ and is defined, according to the normalization condition, by:

$$\kappa = \sqrt{\frac{dT^i}{ds}\frac{dT^i}{ds}} \qquad (109)$$

For a $t$-parameterized curve, the normal unit vector is given by:

$$\mathbf{N} = \frac{\dot{\mathbf{r}}(t) \times (\ddot{\mathbf{r}}(t) \times \dot{\mathbf{r}}(t))}{|\dot{\mathbf{r}}(t)||\ddot{\mathbf{r}}(t) \times \dot{\mathbf{r}}(t)|} \qquad (110)$$

The vector $\mathbf{N}$ is also called the principal normal vector. Based on Eq. 108, this vector is defined only on points of the curve where the curvature $\kappa \neq 0$. Also, since $\mathbf{T}$ is a unit vector then its derivative is orthogonal to it, so the above-stated facts are consistent. We note that Eq. 109 is based on an underlying Cartesian coordinate system. For general curvilinear coordinates, the formula becomes:

$$\kappa = \sqrt{g_{ij}\frac{\delta T^i}{\delta s}\frac{\delta T^j}{\delta s}} \qquad (111)$$

where $g_{ij}$ is the space covariant metric tensor and the notation of absolute derivative is in use (see § 7).

The binormal unit vector is defined as:

$$[\mathbf{B}]^i = B^i = \frac{1}{\tau}\left(\kappa T^i + \frac{dN^i}{ds}\right) \qquad (112)$$

which is a linear combination of two vectors both of which are perpendicular to $N^i$ and hence it is perpendicular to $N^i$. In the last equation, the normalization scalar factor $\tau$ is the "torsion" whose sign is chosen to make $\mathbf{T}, \mathbf{N}, \mathbf{B}$ a right handed system satisfying the condition:

$$\underline{\epsilon}_{ijk}T^i N^j B^k = 1 \qquad (113)$$

where $\underline{\epsilon}_{ijk}$ is the covariant absolute permutation tensor for the 3D space. We should also impose the condition $\tau \neq 0$ on Eq. 112. For plane curves, where the torsion vanishes identically (see § 2.3.2), and at the points with $\tau = 0$ of twisted curves, the binormal unit vector $\mathbf{B}$ may be defined geometrically or as the cross product of $\mathbf{T}$ and $\mathbf{N}$, i.e. $\mathbf{B} = \mathbf{T} \times \mathbf{N}$.

## 2.2 Mathematical Description of Curves

For a $t$-parameterized curve, the binormal vector is given by:

$$\mathbf{B} = \frac{\dot{\mathbf{r}}(t) \times \ddot{\mathbf{r}}(t)}{|\dot{\mathbf{r}}(t) \times \ddot{\mathbf{r}}(t)|} \qquad (114)$$

Inline with making $\mathbf{T}, \mathbf{N}, \mathbf{B}$ a right handed system, there is a geometric significance for the sign of the torsion as it affects the orientation of the space curve. It should be remarked that some authors reverse the sign in the definition of $\tau$ and this reversal affects the signs in the forthcoming Frenet-Serret formulae (see § 2.5). The convention that we follow in this book may have certain advantages.

At any point on the space curve, the triad $\mathbf{T}, \mathbf{N}, \mathbf{B}$ represent a mutually perpendicular right handed system fulfilling the condition:

$$B^i = [\mathbf{T} \times \mathbf{N}]^i = \underline{\epsilon}^{ijk} T_j N_k \qquad (115)$$

where $\underline{\epsilon}^{ijk}$ is the contravariant absolute permutation tensor for the 3D space. Since the vectors in the triad $\mathbf{T}, \mathbf{N}, \mathbf{B}$ are mutually perpendicular, they satisfy the conditions:

$$\mathbf{T} \cdot \mathbf{N} = \mathbf{T} \cdot \mathbf{B} = \mathbf{N} \cdot \mathbf{B} = 0 \qquad (116)$$

Moreover, because they are unit vectors they also satisfy the conditions:

$$\mathbf{T} \cdot \mathbf{T} = \mathbf{N} \cdot \mathbf{N} = \mathbf{B} \cdot \mathbf{B} = 1 \qquad (117)$$

It should be remarked that the triad $\mathbf{T}, \mathbf{N}, \mathbf{B}$ form what is called the Frenet frame which represents a set of orthonormal basis vectors for the embedding space. This frame serves as a basis for a moving orthogonal coordinate system on the points of the curve. The Frenet frame varies in general as it moves along the curve and hence it is a function of the position on the curve. The triad $\mathbf{T}, \mathbf{N}, \mathbf{B}$ may also be called the Frenet trihedron or the moving trihedron of the curve. The Frenet frame can suffer from problems or become undefined, e.g. at non-regular points where $\mathbf{T}$ is undefined or at inflection points where $\frac{d\mathbf{T}}{ds} = \mathbf{0}$.

The tangent line of a curve $C$ at a given point $P$ on the curve is a straight line passing through $P$ and is parallel to the tangent vector, $\mathbf{T}$, of $C$ at $P$. The principal normal line of a curve $C$ at a given point $P$ on the curve is a straight line passing through $P$ and is parallel to the principal normal vector, $\mathbf{N}$, of $C$ at $P$. The binormal line of a curve $C$ at a given point $P$ on the curve is a straight line passing through $P$ and is parallel to the binormal vector, $\mathbf{B}$, of $C$ at $P$. As a consequence, the equations of the three lines can be given by the following generic form:

$$\mathbf{r} = \mathbf{r}_P + k \mathbf{V}_P \qquad (118)$$

where $\mathbf{r}$ is the position vector of an arbitrary point on the line, $\mathbf{r}_P$ is the position vector of the point $P$, $k$ is a real variable ($-\infty < k < \infty$) and the vector $\mathbf{V}_P$ is the vector corresponding to the particular line, that is $\mathbf{V}_P \equiv \mathbf{T}$ for the tangent line, $\mathbf{V}_P \equiv \mathbf{N}$ for the principal normal line, and $\mathbf{V}_P \equiv \mathbf{B}$ for the binormal line.

## 2.2 Mathematical Description of Curves

At any point $P$ on a space curve where the Frenet frame is defined, the triad $\mathbf{T}, \mathbf{N}, \mathbf{B}$ define three mutually perpendicular planes where each one of these planes passes through the point $P$ and is formed by a linear combination of two of these vectors in turn. These planes are: the "osculating plane" which is the span of $\mathbf{T}$ and $\mathbf{N}$, the "rectifying plane" which is the span of $\mathbf{T}$ and $\mathbf{B}$, and the "normal plane" which is the span of $\mathbf{N}$ and $\mathbf{B}$ and is orthogonal to the curve at $P$ (Fig. 24). As a result, the equations of the three planes can be given by the following generic form:

$$(\mathbf{r} - \mathbf{r}_P) \cdot \mathbf{V}_P = 0 \qquad (119)$$

where $\mathbf{r}$ is the position vector of an arbitrary point on the plane, $\mathbf{r}_P$ is the position vector of the point $P$, and where for each plane the vector $\mathbf{V}_P$ is the perpendicular vector to the plane at $P$, that is $\mathbf{V}_P \equiv \mathbf{B}$ for the osculating plane, $\mathbf{V}_P \equiv \mathbf{N}$ for the rectifying plane, and $\mathbf{V}_P \equiv \mathbf{T}$ for the normal plane. Following the style of the definition of the tangent line of a curve as the limit of the secant line (see § 2.1), the osculating plane may also be defined as the limiting position of a plane passing through $P$ and two other points on the curve as the two points converge simultaneously along the curve to $P$.

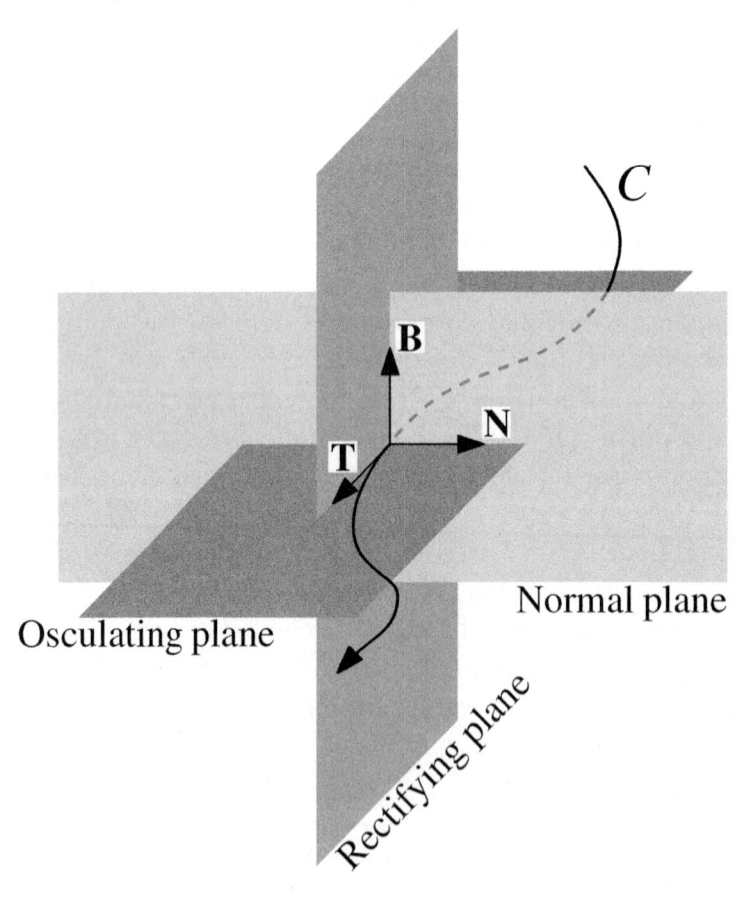

Figure 24: Frenet surfaces and basis vectors at a point on a space curve $C$.

It is noteworthy that the positive sense of a parameterized curve, which corresponds to the direction in which the parameter increases and hence defines the orientation of the curve, can be determined in two opposite ways. While the sense of the tangent **T** and the binormal **B** is dependent on the curve orientation and hence they are in opposite directions in these two ways, the principal normal **N** is the same as it remains parallel to the normal plane in the direction in which the curve is turning. This is consistent with the fact that the triad **T**, **N**, **B** form a right handed system.

## 2.3 Curvature and Torsion of Space Curves

The curvature and torsion of space curves may also be called the first and second curvatures respectively, and hence a twisted curve with non-vanishing curvature and non-vanishing torsion is described as double-curvature curve. The expression $\sqrt{(ds_\mathbf{T})^2 + (ds_\mathbf{B})^2}$, where $ds_\mathbf{T}$ and $ds_\mathbf{B}$ are respectively the lengths of the line element components in the tangent and binormal directions, may be described as the total or the third curvature of the curve. The equation of Lancret states that:

$$(ds_\mathbf{N})^2 = (ds_\mathbf{T})^2 + (ds_\mathbf{B})^2 \tag{120}$$

where $ds_\mathbf{N}$ is the length of the line element component in the principal normal direction. We note that the term "total curvature" is also used for surfaces (see § 4.4 and 4.8) but the meaning is obviously different.

According to the fundamental theorem of space curves in differential geometry, a space curve is completely determined by its curvature and torsion. More technically, given a real interval $I \subseteq \mathbb{R}$ and two differentiable real functions: $\kappa(s) > 0$ and $\tau(s)$ where $s \in I$, there is a uniquely defined parameterized regular space curve $C(s)$: $I \to \mathbb{R}^3$ of class $C^2$ with $\kappa(s)$ and $\tau(s)$ being the curvature and torsion of $C$ respectively and $s$ is its arc length. Hence, any other curve meeting these conditions will be different from $C$ only by a rigid motion transformation (i.e. translation and rotation) which determines its position and orientation in space. On the other hand, any curve with the above-described properties possesses uniquely defined $\kappa(s)$ and $\tau(s)$. As a consequence of the last statements, the fundamental theorem of space curves provides the existence and uniqueness conditions for curves. We note in this context that in rigid motion transformation, which may also be called Euclidean motion, the distance between any two points on the image is the same as the distance between the corresponding points on the inverse image. Hence, rigid motion transformation is a form of isometric mapping.

The equations: $\kappa = \kappa(s)$ and $\tau = \tau(s)$, where $s$ is the arc length, are called the intrinsic or natural equations of the curve. The curvature and torsion are invariants of the space curve and hence they do not depend in magnitude on the employed coordinate system or the type of parameterization. While the curvature is always non-negative ($\kappa \geq 0$), as it represents the magnitude of a vector according to the above-stated definition (see Eqs. 108, 109 and 111), the torsion can be negative as well as zero or positive. It is worth mentioning that some authors define the curvature vector (see § 4.1) and the principal normal vector of space curves in such a way that it is possible for the curvature to be negative.

## 2.3.1 Curvature

The following are some examples of the curvature and torsion of a number of commonly-occurring simple curves:
1. Straight line: $\kappa = 0$ and $\tau = 0$.
2. Circle of radius $R$: $\kappa = \frac{1}{R}$ and $\tau = 0$. Hence, the radius of curvature (see § 2.3.1) of a circle is its own radius
3. Helix parameterized by $\mathbf{r}(t) = (a\cos(t), a\sin(t), bt)$: $\kappa = \frac{a}{a^2+b^2}$ and $\tau = \frac{b}{a^2+b^2}$. It is worth noting that a space curve of class $C^3$ with non-vanishing curvature is a helix *iff* the ratio of its torsion to curvature is constant.

In the above three examples, the curvature and torsion are constant along the whole curve. However, in general the curvature and torsion of space curves are position dependent and hence they vary from point to point.

Following the example of 2D surfaces, a 1D inhabitant of a space curve can detect all the properties related to the arc length. Hence, the curvature and torsion, $\kappa$ and $\tau$, of the curve are extrinsic properties for such a 1D inhabitant. This fact may be expressed by saying that curves are intrinsically Euclidean, and hence their Riemann-Christoffel curvature tensor vanishes identically and they naturally admit 1D Cartesian systems represented by their natural parameterization of arc length. This should be obvious when considering that any curve can be mapped isometrically to a straight line where both are naturally parameterized by arc length. Another demonstration of their intrinsic 1D nature is represented by the forthcoming Frenet-Serret formulae (see § 2.5).

It is noteworthy that some authors resemble the role of $\kappa$ and $\tau$ in curve theory to the surface curvature tensor $b_{\alpha\beta}$ in surface theory (see § 3.4) and describe $\kappa$ and $\tau$ as the curve theoretic analogues of $b_{\alpha\beta}$ in surface theory. In another context, $\kappa$ and $\tau$ may be compared (non-respectively!) with the first and second fundamental forms of surfaces in their roles in defining the curve and surface in the fundamental theorems of these structures (compare the above with what is coming in § 3.6). The curvature and torsion also play in the Frenet-Serret formulae for space curves a similar role to the role played by the coefficients of the first and second fundamental forms in the Gauss-Weingarten equations for space surfaces (see § 3.9). Another useful remark in this context is that from the first and the last of the Frenet-Serret formulae (see Eqs. 136 and 138), we have:

$$|\kappa\tau| = |\mathbf{T}' \cdot \mathbf{B}'| \tag{121}$$

where the prime stands for derivative with respect to a natural parameter $s$ of the curve.

### 2.3.1 Curvature

The curvature $\kappa$ of a space curve is a measure of how much the curve bends as it progresses in the tangent direction at a particular point. The curvature represents the magnitude of the rate of change of the direction of the tangent vector with respect to the arc length and hence it is a measure for the departure of the curve from the orientation of the straight line passing through that point and oriented in the tangent direction. Consequently, the curvature vanishes identically for straight lines (see § 5.1). In fact, having an identically

## 2.3.1 Curvature

vanishing curvature is a necessary and sufficient condition for a curve of class $C^2$ to be a straight line. From the first of the Frenet-Serret formulae (Eq. 136) and the fact that:

$$(\mathbf{N} \cdot \mathbf{T})' = (0)' = 0 \qquad \Rightarrow \qquad \mathbf{N} \cdot \mathbf{T}' = -\mathbf{N}' \cdot \mathbf{T} \tag{122}$$

which is based on the orthogonality of $\mathbf{N}$ and $\mathbf{T}$ and the product rule of differentiation, the curvature $\kappa$ can be expressed as:

$$\kappa = \mathbf{N} \cdot \mathbf{T}' = -\mathbf{N}' \cdot \mathbf{T} \tag{123}$$

where the prime represents differentiation with respect to the arc length $s$ of the curve. The minus sign in the second equality is consistent with the fact that $\kappa$ is non-negative since the component of $\mathbf{N}'$ in the tangential direction is anti-parallel to $\mathbf{T}$. As for the first equality, $\mathbf{N}$ and $\mathbf{T}'$ are parallel (see Eqs. 108 and 110) and hence the dot product is non-negative as it should be.

The "radius of curvature", which is the radius of the osculating circle (see § 2.6), is defined at each point of a space curve at which $\kappa \neq 0$ as the reciprocal of the curvature, that is:

$$R_\kappa = \frac{1}{\kappa} \tag{124}$$

A different way for introducing these concepts, which is followed by some authors, is to define first the radius of curvature as the reciprocal of the magnitude of the acceleration vector, that is $R_\kappa = \frac{1}{|\mathbf{r}''(s)|}$ where $\mathbf{r}(s)$ is the spatial representation of an $s$-parameterized curve; the curvature is then defined as the reciprocal of the radius of curvature. Hence, the radius of curvature may be described as the reciprocal of the norm of the acceleration vector where acceleration means the second derivative of the spatial representation of the curve with respect to its natural parameter.

There may be an advantage in using the concept of "curvature" as the principal concept instead of "radius of curvature", that is the curvature can be defined at all points of a smooth curve where a tangent vector is defined, including those with vanishing curvature, while the radius of curvature is defined only at those points with non-vanishing curvature.

As indicated above, if $C$ is a space curve of class $C^2$ which is defined on a real interval $I \subseteq \mathbb{R}$ and is parameterized by arc length $s \in I$, that is $C(s) : I \to \mathbb{R}^3$, then the curvature of $C$ at a given point $P$ on the curve is defined by:

$$\kappa = |\mathbf{r}''(s)| \tag{125}$$

where $\mathbf{r}(s)$ is the spatial representation of the curve, the double prime represents the second derivative with respect to $s$, and $\mathbf{r}''$ is evaluated at $P$.

For a space curve represented parametrically by $\mathbf{r}(t)$, where $t$ is a general parameter, we have:

$$\kappa = \frac{|\dot{\mathbf{T}}|}{|\dot{\mathbf{r}}|} = \frac{|\dot{\mathbf{r}} \times \ddot{\mathbf{r}}|}{|\dot{\mathbf{r}}|^3} = \frac{\sqrt{(\dot{\mathbf{r}} \cdot \dot{\mathbf{r}})(\ddot{\mathbf{r}} \cdot \ddot{\mathbf{r}}) - (\dot{\mathbf{r}} \cdot \ddot{\mathbf{r}})^2}}{(\dot{\mathbf{r}} \cdot \dot{\mathbf{r}})^{3/2}} \tag{126}$$

where all the quantities, which are functions of $t$, are evaluated at a given point corresponding to a given value of $t$, and the overdot represents derivative with respect to $t$. It is noteworthy that all surface curves passing through a point $P$ on a surface $S$ and have the same osculating plane at $P$ have identical curvature $\kappa$ at $P$ if the osculating plane is not tangent to $S$ at $P$.

## 2.3.2 Torsion

The torsion $\tau$ represents the rate of change of the osculating plane, and hence it quantifies the twisting, in magnitude and sense, of the space curve out of the plane of curvature and its deviation from being a plane curve (see § 5.2). The torsion therefore vanishes identically for plane curves. In fact, having an identically vanishing torsion is a necessary and sufficient condition for a curve of class $C^2$ to be a plane curve. If $C$ is a space curve of class $C^2$ which is defined on a real interval $I \subseteq \mathbb{R}$ and it is parameterized by arc length $s \in I$, that is $C(s) : I \to \mathbb{R}^3$, then the torsion of $C$ at a given point $P$ on the curve is given by:

$$\tau = \mathbf{N}' \cdot \mathbf{B} \tag{127}$$

where $\mathbf{N}'$ and $\mathbf{B}$ are evaluated at $P$ and the prime represents differentiation with respect to $s$. This equation can be obtained from the second of the Frenet-Serret formulae (Eq. 137) by dot producting both sides with $\mathbf{B}$. The formula may also be given as:

$$\tau = -\mathbf{N} \cdot \mathbf{B}' \tag{128}$$

for the same reason as that given for the alternative formulae of $\kappa$ (see Eq. 123 and related text) or by dot producting both sides of the third of the Frenet-Serret formulae (Eq. 138) with $\mathbf{N}$.

For a space curve represented parametrically by $\mathbf{r}(t)$, where $t$ is a general parameter, we have:

$$\tau = \frac{\dot{\mathbf{r}} \cdot (\ddot{\mathbf{r}} \times \dddot{\mathbf{r}})}{|\dot{\mathbf{r}} \times \ddot{\mathbf{r}}|^2} = \frac{\dot{\mathbf{r}} \cdot (\ddot{\mathbf{r}} \times \dddot{\mathbf{r}})}{(\dot{\mathbf{r}} \times \ddot{\mathbf{r}}) \cdot (\dot{\mathbf{r}} \times \ddot{\mathbf{r}})} = \frac{\dot{\mathbf{r}} \cdot (\ddot{\mathbf{r}} \times \dddot{\mathbf{r}})}{(\dot{\mathbf{r}} \cdot \dot{\mathbf{r}})(\ddot{\mathbf{r}} \cdot \ddot{\mathbf{r}}) - (\dot{\mathbf{r}} \cdot \ddot{\mathbf{r}})^2} \tag{129}$$

where all the quantities, which are functions of $t$, are evaluated at a given point $P$ corresponding to a given value of $t$, and the overdot represents derivative with respect to $t$. The curve should have non-vanishing curvature $\kappa$ at $P$.

For rectangular Cartesian coordinates, the torsion of an $s$-parameterized curve is given in tensor notation by:

$$\tau = \frac{\epsilon_{ijk} x'_i x''_j x'''_k}{\kappa^2} \tag{130}$$

where $\kappa$ is the curvature of the curve as defined previously. The last formula is based on its predecessor. For general curvilinear coordinates, the torsion of an $s$-parameterized curve is given in tensor notation by:

$$\tau = \underline{\epsilon}^{ijk} T_i N_j \frac{\delta N_k}{\delta s} \tag{131}$$

The magnitude of torsion is independent of the nature of the curve parameterization and orientation as determined by the sense of increase of its parameter. It is also invariant under permissible coordinate transformations. Finally, the "radius of torsion" is defined at each point of a space curve for which $\tau \neq 0$ as the absolute value of the reciprocal of the torsion, that is:

$$R_\tau = \left| \frac{1}{\tau} \right| \tag{132}$$

We note that some authors do not take the absolute value and accordingly the radius of torsion can be negative.

## 2.4 Geodesic Torsion

Geodesic torsion, which is also known as the relative torsion, is an attribute of a curve embedded in a surface. The geodesic torsion of a surface curve $C$ at a given point $P$ is the torsion of the geodesic curve (see § 5.7) that passes through $P$ in the tangent direction of $C$ at $P$. As we will see in § 5.7, in the neighborhood of a given point $P$ on a smooth surface and for any specified direction there is one and only one geodesic curve passing through $P$ in that direction.

The geodesic torsion $\tau_g$ of a surface curve represented spatially by $\mathbf{r}(s)$ is given by the following scalar triple product:

$$\tau_g = \mathbf{n} \cdot (\mathbf{n}' \times \mathbf{r}') \tag{133}$$

where $\mathbf{n}$ is the unit normal vector to the surface, the primes represent differentiation with respect to the natural parameter $s$, and all these quantities are evaluated at a given point on the curve corresponding to a given value of $s$.

The geodesic torsion of a curve $C$ at a non-umbilical point $P$ (see § 4.10) is given in terms of the principal curvatures $\kappa_1$ and $\kappa_2$ (see § 4.4) by:

$$\tau_g = (\kappa_1 - \kappa_2) \sin\theta \cos\theta \tag{134}$$

where $\theta$ is the angle between the tangent vector $\mathbf{T}$ to the curve $C$ at $P$ and the first principal direction $\mathbf{d}_1$ (see Darboux frame in § 4.4).

The geodesic torsion of a surface curve $C$ parameterized by arc length $s$ at a given point $P$ is also given in terms of the curve torsion by:

$$\tau_g = \tau - \frac{d\phi}{ds} \tag{135}$$

where $\tau$ is the torsion of $C$ at $P$, and $\phi$ is the angle between the unit normal vector $\mathbf{n}$ to the surface and the principal normal vector $\mathbf{N}$ of $C$ at $P$, i.e. $\phi = \arccos(\mathbf{n} \cdot \mathbf{N})$. Also, the curve $C$ should not be asymptotic (see § 5.9). This formula (Eq. 135) which is known as the Bonnet formula, demonstrates that when $\mathbf{n}$ and $\mathbf{N}$ are collinear along the curve, the geodesic torsion and the torsion are equal (i.e. $\tau_g = \tau$). As we will see in § 5.7, when $\mathbf{n}$ and $\mathbf{N}$ are collinear, the geodesic component of the curvature vector (see § 4.1) will vanish. In this case, the geodesic curvature (see § 4.3) will vanish and the curve becomes a geodesic. So in brief, on a geodesic curve we have: $\tau_g = \tau$ which is consistent with the above statement at the beginning of this section.

The geodesic torsion of a surface curve $C$ at a given point $P$ is zero *iff* $C$ is tangent to a line of curvature at $P$ (see § 5.8). Hence, on a line of curvature the geodesic torsion vanishes identically. This can be seen from Eq. 134 where either $\sin\theta$ or $\cos\theta$ vanishes. The geodesic torsions of two orthogonal surface curves at their point of intersection are equal in magnitude and opposite in sign.

## 2.5 Relationship between Curve Basis Vectors and their Derivatives

The three basis vectors $\mathbf{T}, \mathbf{N}, \mathbf{B}$ of a space curve are connected to their derivatives by the Frenet-Serret formulae which are given in rectangular Cartesian coordinates by:

$$\frac{dT^i}{ds} = \kappa N^i \tag{136}$$

$$\frac{dN^i}{ds} = \tau B^i - \kappa T^i \tag{137}$$

$$\frac{dB^i}{ds} = -\tau N^i \tag{138}$$

As indicated previously (see § 2.2), the sign of the terms involving $\tau$ depends on the convention about the torsion and hence these equations differ between different authors. The above equations are also known as Frenet formulae.

The Frenet-Serret formulae can be cast in the following matrix form using symbolic notation:

$$\begin{bmatrix} \mathbf{T}' \\ \mathbf{N}' \\ \mathbf{B}' \end{bmatrix} = \begin{bmatrix} 0 & \kappa & 0 \\ -\kappa & 0 & \tau \\ 0 & -\tau & 0 \end{bmatrix} \begin{bmatrix} \mathbf{T} \\ \mathbf{N} \\ \mathbf{B} \end{bmatrix} \tag{139}$$

where all the quantities in this equation are functions of arc length $s$ and the prime represents derivative with respect to $s$. As seen, the coefficient matrix of this system is anti-symmetric.

The Frenet-Serret formulae can also be given in the following form:

$$\mathbf{T}' = \mathbf{d} \times \mathbf{T} \tag{140}$$

$$\mathbf{N}' = \mathbf{d} \times \mathbf{N} \tag{141}$$

$$\mathbf{B}' = \mathbf{d} \times \mathbf{B} \tag{142}$$

where $\mathbf{d}$ is the "Darboux vector" which is given by:

$$\mathbf{d} = \tau \mathbf{T} + \kappa \mathbf{B} \tag{143}$$

This form of the Frenet-Serret formulae is more memorable apart from the expression of $\mathbf{d}$. We note that some authors define $\mathbf{d}$ as a scalar multiple of what is given in Eq. 143.

## 2.6 Osculating Circle and Sphere

The above three equations (i.e. Eqs. 140-142) may be merged in a single equation as:

$$(\mathbf{T'}, \mathbf{N'}, \mathbf{B'}) = \mathbf{d} \times (\mathbf{T}, \mathbf{N}, \mathbf{B}) \tag{144}$$

In general curvilinear coordinates, the Frenet-Serret formulae are given in terms of the absolute derivatives of the three vectors by:

$$\frac{\delta T^i}{\delta s} = \frac{dT^i}{ds} + \Gamma^i_{jk} T^j \frac{dx^k}{ds} = \kappa N^i \tag{145}$$

$$\frac{\delta N^i}{\delta s} = \frac{dN^i}{ds} + \Gamma^i_{jk} N^j \frac{dx^k}{ds} = \tau B^i - \kappa T^i \tag{146}$$

$$\frac{\delta B^i}{\delta s} = \frac{dB^i}{ds} + \Gamma^i_{jk} B^j \frac{dx^k}{ds} = -\tau N^i \tag{147}$$

where the indexed $x$ represent general spatial coordinates and $s$ is a natural parameter while the other symbols are as defined earlier.

According to the fundamental theorem of space curves, which is outlined previously in § 2.3, a curve does exist and it is unique *iff* its curvature and torsion as functions of arc length are given. Now, it is natural to expect that such a solution can be obtained from the system of differential equations given by the Frenet-Serret formulae. However, such a solution cannot be obtained in general by direct integration of these equations. More elaborate methods (e.g. methods based on the Riccati equation for reducing a system of simultaneous differential equations to a first order differential equation) may be used to obtain the solution. Nevertheless, a solution can be obtained by direct integration of the Frenet-Serret formulae for plane curves (see § 5.2) where the torsion vanishes identically. A solution by direct integration of the Frenet-Serret formulae can also be obtained in other simple cases such as when the curvature and torsion are constants.

## 2.6 Osculating Circle and Sphere

At any point $P$ with non-zero curvature of a smooth space curve $C$, an "osculating circle" (Fig. 25), which may also be called the circle of curvature or the kissing circle, can be defined where this circle is characterized by:
1. It is tangent to $C$ at $P$, i.e. the circle and the curve have a common tangent vector at $P$.
2. It lies in the osculating plane of $C$ at $P$.
3. Its radius $R_\kappa$ is equal to $\frac{1}{\kappa}$ where $\kappa$ is the curvature of $C$ at $P$.
4. Its center $C_c$ is located at $\mathbf{r}_C$ which is given by:

$$\mathbf{r}_C = \mathbf{r}_P + \frac{1}{\kappa}\mathbf{N} = \mathbf{r}_P + R_\kappa \mathbf{N} \tag{148}$$

where $\mathbf{r}_P$ is the position vector of $P$ and $\mathbf{N}$ is the principal normal vector of $C$ at $P$. The center of curvature of a curve at a point on the curve is defined as the center of the osculating circle at that point, as given above. If the curve $C$ is a circle, then the center

## 2.6 Osculating Circle and Sphere

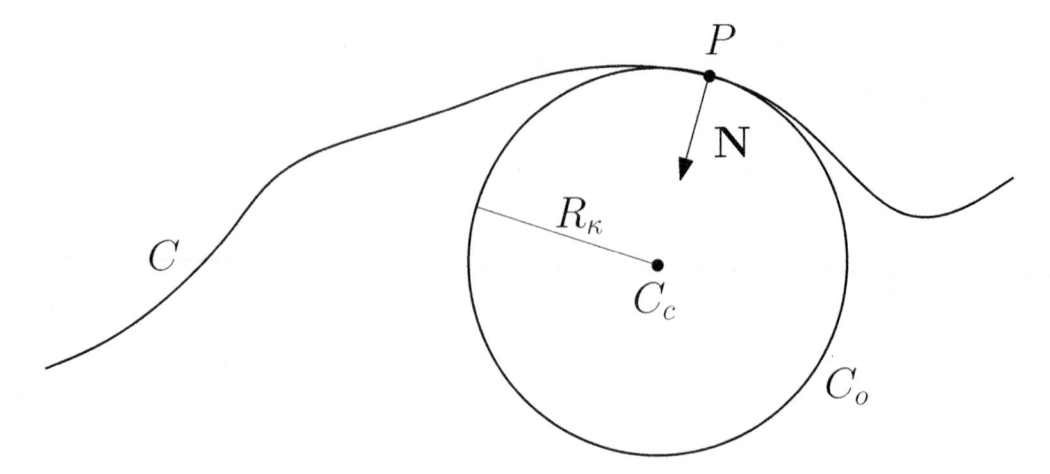

Figure 25: The osculating circle $C_o$ of a space curve $C$ at point $P$ with the principal normal vector **N**, center of curvature $C_c$ and radius of curvature $R_\kappa$.

of curvature at any point is the center of the circle itself, so the circle is its own osculating circle.

The osculating circle provides a good approximation to the curve in the neighborhood of its points where the osculating circle is defined. Following the manner of defining the tangent line to a curve as a limit of the secant line (see § 2.1), the osculating circle to a curve at a given point $P$ may be defined geometrically as the limit of a circle passing through $P$ and two other points on the curve as these two points converge to $P$ while staying on the curve (Fig 26). It should be remarked that in some cases the osculating circle and its parameters may be defined geometrically but not analytically when the second derivative of the curve at the given point is not properly defined to determine the radius of curvature (see Eq. § 125).

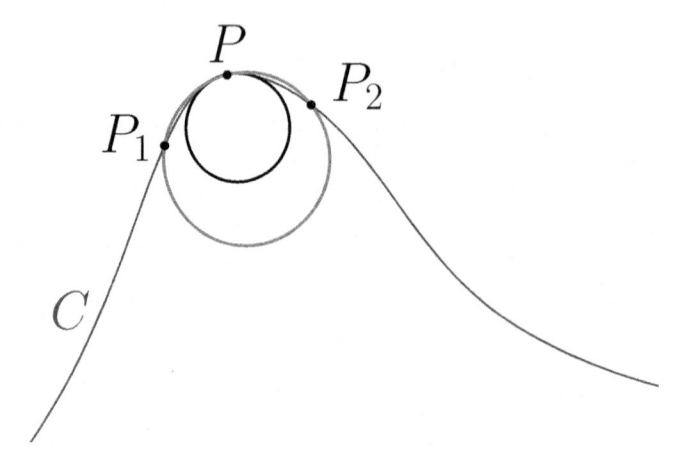

Figure 26: The osculating circle (small) of a curve $C$ at a point $P$ as a limit of another circle (big) passing through $P$ and two other points, $P_1$ and $P_2$.

Following the manner of defining the osculating circle as a limit, the "osculating sphere" of a curve $C$ at a given point $P$ may be defined similarly as the limit of a sphere passing through $P$ and three neighboring points on the curve as these three points converge to $P$. The position of the center $C_S$ of the osculating sphere at $P$, which is called the center of spherical curvature of $C$ at $P$, is given by:

$$\mathbf{r}_S = \mathbf{r}_P + \frac{1}{\kappa}\mathbf{N} - \frac{\kappa'}{\tau\kappa^2}\mathbf{B} = \mathbf{r}_P + R_\kappa \mathbf{N} + \operatorname{sgn}(\tau) R_\tau R'_\kappa \mathbf{B} \tag{149}$$

where $\mathbf{r}_S$ and $\mathbf{r}_P$ are the position vectors of $C_S$ and $P$, $\mathbf{B}$ and $\mathbf{N}$ are the binormal and principal normal vectors, $\kappa$ and $\tau$ are the curvature and torsion, $R_\kappa$ and $R_\tau$ are the radii of curvature and torsion, $\operatorname{sgn}(\tau)$ is the sign function of $\tau(s)$, and the prime represents derivative with respect to a natural parameter $s$ of $C$. All these quantities belong to $C$ at $P$ which should have non-vanishing curvature and torsion, i.e. $\kappa, \tau \neq 0$. From Eq. 149, it can be seen that the radius of the osculating sphere is given by:

$$|\mathbf{r}_S - \mathbf{r}_P| = \sqrt{R_\kappa^2 + (R_\tau R'_\kappa)^2} \tag{150}$$

## 2.7 Parallelism and Parallel Propagation

In flat spaces, parallelism is an absolute property as it is defined without reference to a peripheral object. However, in Riemannian spaces the idea of parallelism is defined in reference to a prescribed curve and hence it is different from the idea of parallelism in the Euclidean sense. A vector field $A^\alpha$ is described as being parallel along the surface curve $u^\beta = u^\beta(t)$ *iff* its absolute derivative (see § 7) along the curve vanishes, that is:

$$\frac{\delta A^\alpha}{\delta t} \equiv A^\alpha_{;\beta} \frac{du^\beta}{dt} \equiv \frac{dA^\alpha}{dt} + \Gamma^\alpha_{\beta\gamma} A^\gamma \frac{du^\beta}{dt} = 0 \tag{151}$$

This means that the sufficient and necessary condition for a vector field to be parallel along a surface curve is that the covariant derivative of the field is normal to the surface.

All the vectors of a field of parallel vectors have the same constant magnitude. A field of absolutely parallel unit vectors on a surface do exist *iff* there is an isometric correspondence between the plane and the surface. When two surfaces are tangent to each other along a given curve $C$, then a vector field which is parallel along $C$ with respect to one of these surfaces will also be parallel along $C$ with respect to the other surface. When two non-trivial vectors experience parallel propagation along a particular curve the angle between them stays constant.

As a consequence of the definition of parallelism in Riemannian spaces and the previous statements, we have:
1. Parallel propagation is path dependent. Hence, a surface vector field parallelly propagated along a given curve between two points $P_1$ and $P_2$ on the curve does not necessarily coincide with another vector field parallelly propagated along another curve connecting $P_1$ and $P_2$.
2. Since parallel propagation depends on the path of propagation, then given two points $P_1$ and $P_2$ on a surface, the vector obtained at $P_2$ by parallel propagation of a vector from $P_1$ along a given surface curve $C$ connecting $P_1$ to $P_2$ depends on the curve $C$.

3. Starting from a given point $P$ on a closed surface curve $C$ enclosing a simply connected region (see § 3.1) on the surface, parallel propagation of a vector field around $C$ starting from $P$ does not necessarily result in the same vector field when arriving at $P$. We note that the angle between the initial and final vectors in this situation is a measure of the Gaussian curvature on the surface (see § 4.5).
4. If $C : I \to S$ is a regular curve on a surface $S$ defined on the interval $I \subseteq \mathbb{R}$, and $\mathbf{v}_1$ and $\mathbf{v}_2$ are parallel vector fields over $C$, then the dot product $\mathbf{v}_1 \cdot \mathbf{v}_2$ which is associated with the metric tensor, the norm of the vector fields $|\mathbf{v}_1|$ and $|\mathbf{v}_2|$, and the angle between $\mathbf{v}_1$ and $\mathbf{v}_2$ are constants.

## 2.8 Exercises

2.1 State the technical definition of space curve outlining the difference between a curve and its trace.

2.2 What is the most common way of defining space curves mathematically? Give an example from simple curves like circle and ellipse.

2.3 Make a clear distinction between general and natural parameterization of space curves.

2.4 State a mathematical condition for a space curve to be parameterized naturally.

2.5 What is the relation between two natural parameters of a given space curve?

2.6 The following equation: $\mathbf{r}(t) = \left(\frac{t}{2}, -\frac{t}{2}, \frac{t}{\sqrt{2}}\right)$ is a parametric representation of a space curve. Is $t$ a natural parameter or not? Justify your answer.

2.7 Prove that two natural parameters, $s$ and $š$, of a curve are related by the equation $š = \pm s + c$ where $c$ is a real constant.

2.8 Using tensor notation, write down the equation of the tangent vector to a surface curve represented parametrically by $C(u^1(t), u^2(t))$ where $t$ is a general parameter.

2.9 What is the meaning of having "regular curve at a specific point"? What "regular curve" means?

2.10 Prove that the parametric representation: $\mathbf{r}(t) = (t^2, e^t, t+1)$ of a space curve is regular for all $t$.

2.11 State the condition for a vector to be tangent to a regular surface at a given point on the surface.

2.12 Find the unit tangent vector, $\mathbf{T}(t)$, for a space curve represented by: $\mathbf{r}(t) = (t^2, t, \sin t)$.

2.13 Find the arc length of a space curve given by: $\mathbf{r}(t) = (5t, 7\cosh t, 2\sinh t)$ for $1 \leq t \leq 4$.

2.14 Find the curvature, as a function of $t$, of a space curve represented by: $\mathbf{r}(t) = (\cos t - 1, \sin t + t, t^2)$.

2.15 Define "periodic curve" giving two common examples other than those given in the text.

2.16 Should a periodic curve be a plane curve?

2.17 Should a periodic continuous curve be a closed curve? Should a periodic smooth curve be a closed curve?

2.18 A plane curve called *cissoid of Diocles* is given in polar coordinates by: $\rho = 2\sin\phi \tan\phi$. Find the equation of the curve in a rectangular Cartesian coordinate system.

2.19 Sketch the curve of the previous question for $0 \leq |\phi| \leq \frac{\pi}{4}$ (notice the two branches).

## 2.8 Exercises

2.20 Show that for a curve represented spatially by **r** and parameterized naturally by $s$ and generally by $t$, the relation between $s$ and $t$ is given by: $\left|\frac{ds}{dt}\right| = \left|\frac{d\mathbf{r}}{dt}\right|$.

2.21 Define Frenet frame with a simple sketch showing the basis vectors at a given point on an arbitrary space curve.

2.22 Write down the mathematical equations of the unit vectors **T**, **N**, **B** for a curve parameterized by a general parameter $t$ and a natural parameter $s$.

2.23 State the mathematical definition of the curvature $\kappa$ and the torsion $\tau$ of an $s$-parameterized space curve.

2.24 Show that the sufficient and necessary condition for a space curve to be a plane curve is that its torsion vanishes identically.

2.25 What are the curvature and torsion of (a) a straight line (b) a circle with radius $R = 3.2$ (c) a curve parameterized by: $\mathbf{r}(t) = (3\cos(t), 3\sin(t), 1.9t)$?

2.26 Find the torsion, as a function of $t$, of (a) a curve represented by: $\mathbf{r}(t) = (\sin t + t, \cos t - 3, t + 2)$ (b) a curve represented by: $\mathbf{r}(t) = (t, 2t^2, t^3)$ (c) a curve represented by: $\mathbf{r}(t) = (at, bt^3, ct^2)$ where $a, b, c$ are non-vanishing real constants.

2.27 What are the mathematical conditions that represent the fact that the vectors **T**, **N**, **B** are mutually orthogonal and of unit length?

2.28 Prove the theorem of Lancret which states that a space curve of class $C^3$ with non-vanishing curvature is a helix *iff* the ratio of its torsion to its curvature is constant along the curve.

2.29 Show that if two space curves, which have an injective association, possess parallel tangent vectors at their corresponding points, then their normal and binormal vectors at these points are parallel as well.

2.30 Prove Eq. 121 (i.e. $|\kappa\tau| = |\mathbf{T}' \cdot \mathbf{B}'|$).

2.31 Give a parametric representation of a circular helix using a natural parameter $s$.

2.32 For a plane curve in a 3D space given by the equation: $y = 2x^2 - x + 3$ $(0 \le x \le 10)$, find the equations of the osculating, normal and rectifying planes at point $(1, 4)$.

2.33 For a space curve parameterized as: $(x, y, z) = (t, t^3, 3t^2)$, find the equations of the tangent, principal normal and binormal lines passing through the point $(1, 1, 3)$ on the curve.

2.34 Give an example of a non-planar space curve whose principal normal vectors at all points of the curve are parallel to a particular plane.

2.35 For a space curve parameterized as: $(x, y, z) = (\cos t, \sin t, 5t)$, find the equations of the osculating, rectifying and normal planes at the point on the curve with $t = 1.3$.

2.36 Which of the three vectors **T**, **N**, **B** is not affected by reversing the sense of traversing the space curve?

2.37 What are the principal normal vector **N** and the binormal vector **B** of a helix represented by: $\mathbf{r}(t) = (\cos 3t, \sin 3t, 3t)$?

2.38 Find the three vectors **T**, **N**, **B** along a curve represented by: $\mathbf{r}(t) = (2t-3, t^3+t, 5-t^2)$.

2.39 Make a simple sketch for the osculating, rectifying and normal planes at a point on an arbitrary space curve. Use a computer graphic package if convenient.

2.40 Write down the equation of Lancret related to the third curvature of space curves and discuss its significance.

## 2.8 Exercises

2.41 Discuss, in detail, the fundamental theorem of space curves in differential geometry outlining its application and significance.

2.42 Given that the curvature of a plane curve is given by: $\kappa = \frac{1}{3s+5}$ where $s > 0$ is a natural parameter, find the equation of this curve.

2.43 Write down the Frenet-Serret formulae of a space curve in a rectangular Cartesian coordinate system explaining all the symbols involved.

2.44 By integrating the Frenet-Serret formulae, obtain the solution of a space curve with $\kappa = a$ and $\tau = b$ where $a, b > 0$ are real constants.

2.45 Identify the type of the curve in the previous question.

2.46 What are the "intrinsic" or "natural" equations of a curve?

2.47 Find the intrinsic equations of the catenary defined by Eqs. 28-29.

2.48 Find the parametric representation of a curve with the following intrinsic equations: $\kappa = \sqrt{c/s}$ and $\tau = 0$ where $c > 0$ is a real constant and $s > 0$ is a natural parameter.

2.49 Prove that the curvature of a $t$-parameterized space curve is given by: $\kappa = \frac{|\dot{\mathbf{r}} \times \ddot{\mathbf{r}}|}{|\dot{\mathbf{r}}|^3}$.

2.50 Prove that the torsion of a $t$-parameterized space curve is given by: $\tau = \frac{\dot{\mathbf{r}} \cdot (\ddot{\mathbf{r}} \times \dddot{\mathbf{r}})}{|\dot{\mathbf{r}} \times \ddot{\mathbf{r}}|^2}$.

2.51 Which of the two main curve parameters, $\kappa$ and $\tau$, is necessarily non-negative and why?

2.52 Show that along an $s$-parameterized curve $\mathbf{r}(s)$, the following relation holds true: $\mathbf{r}' \cdot (\mathbf{r}'' \times \mathbf{r}''') = \tau \kappa^2$.

2.53 Can we obtain the curvature and torsion of circle as a special case of the curvature and torsion of helix? If yes, how? Is this consistent with the definition of helix as given in § 1.4.1 (see Eqs. 4-6 and Fig. 2)?

2.54 Find the curvature and torsion of a space curve represented by: $\mathbf{r}(t) = (t^3, t+1, -t^2)$ at the point with $t = 2.4$.

2.55 Give an example of a space curve whose curvature and torsion are variables.

2.56 Investigate the relation between the curvature and torsion at corresponding points of two space curves which are mirror-reflection of each other with respect to a given plane.

2.57 Investigate the relation between the curvature and torsion at corresponding points of two space curves which are symmetric with respect to a given point.

2.58 Discuss, in detail, the concept of "1D inhabitant" in the context of classifying the properties of space curves.

2.59 Establish a correspondence between the two main parameters of space curve (i.e. curvature and torsion) and the first and second fundamental forms of space surface.

2.60 Discuss the resemblance between $\kappa$ and $\tau$ of space curve and the curvature tensor of space surface.

2.61 State, in words, the mathematical relation: $\tau = \epsilon^{ijk} T_i N_j \frac{\delta N_k}{\delta s}$ using technical terms for all the notations and symbols used in this equation.

2.62 What is the significance of the curvature and torsion of space curves as measures of their variation in the embedding space?

2.63 What is the relation between the curvature and the radius of curvature of a space curve? Is there an advantage in using one of these or the other as the main concept?

## 2.8 Exercises

2.64 Outline the physical significance of the relation: $\kappa = |\mathbf{r}''|$.

2.65 Express $\kappa$ in terms of $\mathbf{r}$ and its first and second derivatives where $\mathbf{r}$ is a spatial representation of a curve parameterized by a general parameter $t$.

2.66 Express $\tau$ in terms of $\mathbf{N}'$ and $\mathbf{B}$ of a naturally parameterized curve.

2.67 Express $\tau$ in terms of $\mathbf{r}$ and its first, second and third derivatives where $\mathbf{r}$ is a spatial representation of a curve parameterized by a general parameter $t$.

2.68 What is the relation between the torsion and the radius of torsion of a space curve?

2.69 Define geodesic torsion in words and state its mathematical relation to $\mathbf{r}$ and $\mathbf{n}$ and their derivatives.

2.70 What is the significance of the relation: $\tau_g = \tau - \frac{d\phi}{ds}$ and what the symbols in this relation mean?

2.71 What is the condition for the torsion and the geodesic torsion of a space curve to be equal?

2.72 Prove that along a sufficiently smooth curve represented by $\mathbf{r}(t)$, the vector $\ddot{\mathbf{r}}$ at a given point $P$ on the curve is parallel to the osculating plane at $P$.

2.73 Obtain the equation of the osculating plane of a curve represented parametrically by: $\mathbf{r}(t) = (3\cos t, 2\sin t, \cos t + 5\sin t)$ at a general point on the curve.

2.74 Derive the second of the Frenet-Serret formulae using the other two formulae.

2.75 Define Darboux vector $\mathbf{d}$ and hence verify that the relations given by Eqs. 140-142 are valid.

2.76 Explain all the symbols and notations used in the following relation: $\tau = \frac{\dot{\mathbf{r}} \cdot (\ddot{\mathbf{r}} \times \dddot{\mathbf{r}})}{(\dot{\mathbf{r}} \cdot \dot{\mathbf{r}})(\ddot{\mathbf{r}} \cdot \ddot{\mathbf{r}}) - (\dot{\mathbf{r}} \cdot \ddot{\mathbf{r}})^2}$.

2.77 Prove that for a curve with helical shape, the Darboux vector is constant along the curve.

2.78 State the Frenet-Serret formulae in a single equation using the Darboux vector.

2.79 Write down the Frenet-Serret formulae assuming a general curvilinear coordinate system.

2.80 Give a brief explanation of why the solution of a space curve cannot be obtained in general by a direct integration of the Frenet-Serret equations.

2.81 Give a brief definition of the osculating circle and the osculating sphere of a space curve.

2.82 How the osculating circle and the osculating sphere of a space curve can be defined using the concept of limit?

2.83 Prove that for a given space curve $C$, the binormal lines of $C$ and the tangent lines to the locus of the centers of spherical curvature of $C$ are parallel at their corresponding points.

2.84 Derive an expression for the position of the center of curvature of a $t$-parameterized twisted curve represented spatially by $\mathbf{r}(t)$ in terms of $\mathbf{r}$ and its derivatives.

2.85 Find the spatial coordinates of the center of the osculating circle of a space curve represented by: $\mathbf{r}(t) = (t, \sqrt{t}, t^2)$ at the point with $t = 2.6$.

2.86 What is the relation between the osculating circle and the osculating plane at a given point of a space curve?

2.87 Explain all the symbols involved in the equation: $\mathbf{r}_S = \mathbf{r}_P + R_\kappa \mathbf{N} + \operatorname{sgn}(\tau) R_\tau R'_\kappa \mathbf{B}$ which is related to the center of the osculating sphere.

## 2.8 Exercises

2.88 Write down the formula for the radius of the osculating sphere explaining all the symbols used.

2.89 What is the difference between the concept of parallelism in Euclidean and non-Euclidean spaces?

2.90 State the mathematical condition for a vector field to be parallel along a given surface curve in terms of its absolute derivative.

2.91 What is the meaning of describing parallel propagation as path dependent? What are the consequences of this dependency?

# Chapter 3
# Surfaces in Space

Here, we examine sufficiently smooth surfaces embedded in a 3D Euclidean space using a Cartesian coordinate system $(x, y, z)$ for the most parts. Some parts are based on a more general Riemannian space with a curvilinear coordinate system.

## 3.1 General Background about Surfaces

A 2D surface embedded in a 3D space may be defined loosely as a set of connected points in the space such that the immediate neighborhood of each point on the surface can be deformed continuously to form a flat disk. Technically, a surface in a 3D manifold is a mapping from a subset of the parameters plane to a 3D space, that is $S : \Omega \to \mathbb{R}^3$, where $\Omega$ is a subset of $\mathbb{R}^2$ plane and $S$ is a sufficiently smooth injective function. Similar conditions may also be imposed to ensure the existence of a tangent plane and a normal vector at each point of the surface. In particular, the condition $\partial_u \mathbf{r} \times \partial_v \mathbf{r} \neq \mathbf{0}$ at all points on the surface is usually imposed to ensure regularity of the surface.

Like space and surface curves, the image of the mapping in $\mathbb{R}^3$ is known as the trace of the surface. For convenience, we use curve and surface in the present book for trace as well as for mapping; the meaning should be obvious from the context. We should remark that the trace of a curve or a surface should not be confused with the trace of a matrix which is the sum of its diagonal elements.

A 2D surface embedded in a 3D space can be defined explicitly by: $z = f(x, y)$, or implicitly by: $F(x, y, z) = 0$, or parametrically by: $x(u^1, u^2), y(u^1, u^2), z(u^1, u^2)$ where $u^1$ and $u^2$ (or $u$ and $v$) are the surface coordinates on the parameters plane, as defined previously (see § 1.4.3), which are mutually independent parameters. By substitution, elimination and algebraic manipulation these forms can be transformed interchangeably.

A coordinate patch of a surface is an injective, bicontinuous, regular, parametric representation of a part of the surface. In more technical terms, a coordinate patch of class $C^n$ ($n > 0$) on a space surface $S$ is a functional mapping of an open set $\Omega$ in the $uv$ parameters plane onto $S$ that satisfies the following conditions:
1. The functional mapping relation is of class $C^n$ over $\Omega$.
2. The mapping is one-to-one and bicontinuous over $\Omega$.
3. $\mathbf{E}_1 \times \mathbf{E}_2 \neq \mathbf{0}$ at any point in $\Omega$ where $\mathbf{E}_1$ and $\mathbf{E}_2$ are the surface basis vectors (see § 1.4.5 and 1.4.3).

As indicated previously, a vector $\mathbf{v}$ is described as a tangent vector to the surface $S$ at a given point $P$ on the surface if there is a regular curve $C$ embedded in $S$ and passing through $P$ such that $\mathbf{v}$ is a tangent to the curve at $P$, i.e. $\mathbf{v} = c\frac{d\mathbf{r}(t)}{dt}$ ($c \in \mathbb{R}$, $c \neq 0$) where $\mathbf{r}$ is a $t$-parameterized position vector representing $C$. The set of all tangent vectors to the surface $S$ at point $P$ forms a tangent plane to $S$ at $P$ (Fig. 27). This set is called

## 3.1 General Background about Surfaces

the tangent space of $S$ at $P$ and it is usually notated with $T_PS$. It is obvious that there exist infinitely many tangent vectors, with varying magnitude and direction, to a smooth surface at its regular points.

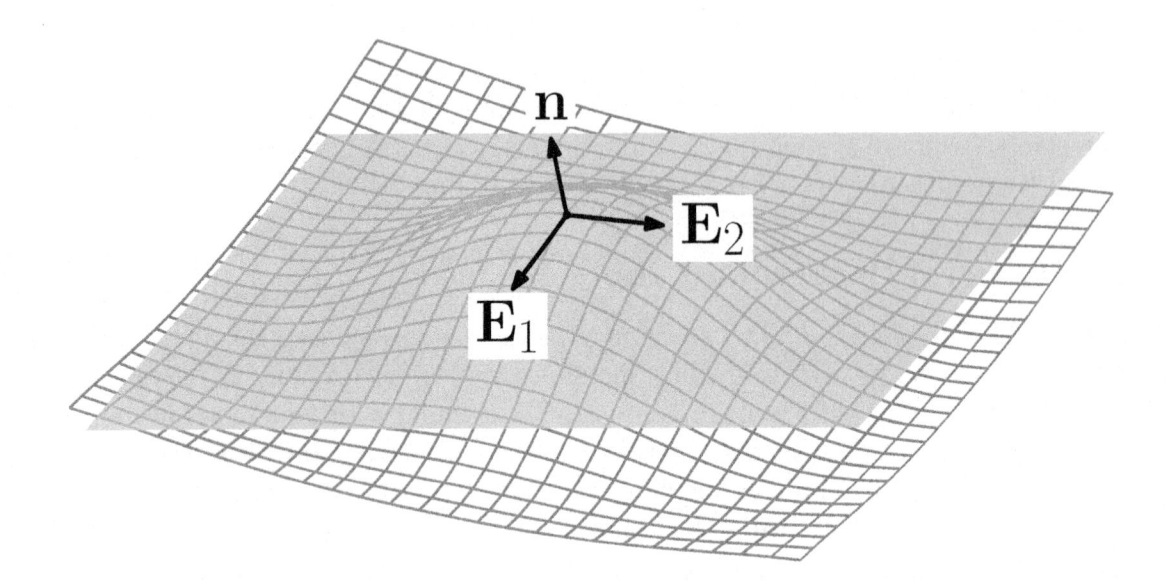

Figure 27: Tangent plane (solid) of a surface (outlined by a grid) at a given point alongside the surface basis vectors, $\mathbf{E}_1$ and $\mathbf{E}_2$, and the unit normal vector, $\mathbf{n}$, at that point.

As we will see (also refer to § 1.4.5), the tangent space of a regular surface at a given point on the surface is the span of the two linearly independent basis vectors defined as:

$$\mathbf{E}_1 = \frac{\partial \mathbf{r}}{\partial u^1} \qquad \mathbf{E}_2 = \frac{\partial \mathbf{r}}{\partial u^2} \qquad (152)$$

where $\mathbf{r}(u^1, u^2)$ is the spatial representation of the surface coordinate curves and $u^1$ and $u^2$ are the surface coordinates, as explained before. The tangent space, therefore, is the plane passing through $P$ and is perpendicular to the vector $\mathbf{E}_1 \times \mathbf{E}_2$. As indicated previously, every vector tangent to a regular surface $S$ at a given point $P$ on $S$ can be expressed as a linear combination of the surface basis vectors $\mathbf{E}_1$ and $\mathbf{E}_2$ at $P$. The reverse is also true, that is every linear combination of $\mathbf{E}_1$ and $\mathbf{E}_2$ at $P$ is a tangent vector to a regular curve embedded in $S$ and passing through $P$ and hence it is a tangent to $S$ at $P$.

Based on the previous statements, we see that the tangent plane of a surface at a given point $P$ on the surface can be given by:

$$\mathbf{r} = \mathbf{r}_P + p\mathbf{E}_1 + q\mathbf{E}_2 \qquad (153)$$

where $\mathbf{r}$ is the position vector of an arbitrary point on the tangent plane, $\mathbf{r}_P$ is the position vector of the point $P$, $p, q \in (-\infty, \infty)$ are real variables, and $\mathbf{E}_1$ and $\mathbf{E}_2$ are the surface basis vectors at $P$. Alternatively, the tangent plane can be expressed in terms of the

## 3.1 General Background about Surfaces

normal vector **n** to the surface at $P$ by:

$$(\mathbf{r} - \mathbf{r}_P) \cdot \mathbf{n} = 0 \tag{154}$$

The tangent space at a specific point $P$ of a surface is a property of the surface at $P$ and hence it is independent of the patch that contains $P$ and the particular parameterization of the surface. For any non-trivial vector **v** which is parallel to the tangent plane of a simple and smooth surface $S$ at a given point $P$ on $S$, there is a curve in $S$ passing through $P$ and represented parametrically by $\mathbf{r}(t)$ such that $\mathbf{v} = c\frac{d\mathbf{r}}{dt}$ where $c \neq 0$ is a real constant. As a result of the last and the previous statements, a non-trivial vector is parallel to the tangent plane of a surface $S$ at a given point $P$ *iff* it is equal to a tangent vector to $S$ at $P$.

The straight line passing through a given point $P$ on a surface $S$ in the direction of the normal vector of $S$ at $P$ is called the normal line to $S$ at $P$. The equation of this normal line is given by:

$$\mathbf{r} = \mathbf{r}_P + k\mathbf{n} \tag{155}$$

where **r** is the position vector of an arbitrary point on the normal line, $\mathbf{r}_P$ is the position vector of the point $P$, $k \in (-\infty, \infty)$ is a real variable, and **n** is the unit normal vector to the surface at $P$. We remark that this normal line should not be confused with the normal line of a surface curve $C$ that passes through $P$ which is usually called the principal normal line of $C$ (see § 2.2). Anyway, the two should be easily distinguished by noticing their affiliation to surface or curve.

As stated before, a surface is regular at a given point $P$ *iff* $\mathbf{E}_1 \times \mathbf{E}_2 \neq \mathbf{0}$ at $P$ where $\mathbf{E}_1 = \partial_u \mathbf{r}$ and $\mathbf{E}_2 = \partial_v \mathbf{r}$ are the tangent vectors to the coordinate curves at $P$. A surface is regular *iff* $\mathbf{E}_1 \times \mathbf{E}_2 \neq \mathbf{0}$ at any point on the surface. A regular curve (see § 2.1) of class $C^n$ on a sufficiently smooth surface is an image of a unique regular plane curve of class $C^n$ in the parameters plane, where in this statement we are considering each connected part of the curve being embedded in a coordinate patch if there is no single patch that contains the entire curve.

A "Monge patch" is a coordinate patch in a 3D space defined by a function in one of the following three forms:

$$\mathbf{r}(u, v) = (f(u, v), u, v) \tag{156}$$
$$\mathbf{r}(u, v) = (u, f(u, v), v) \tag{157}$$
$$\mathbf{r}(u, v) = (u, v, f(u, v)) \tag{158}$$

where $f$ is a differentiable function of the surface coordinates $u$ and $v$. When $f$ is of class $C^n$ then the coordinate patch is of this class.

A **simply connected** region on a surface means that a closed curve contained in the region can be shrunk down continuously onto any point in the region without leaving the region. In more simple terms, it means that the region contains no holes or gaps that separate its parts. A **simple** surface is a continuously deformed plane by compression, stretching and bending.[10] Examples of simple surface are cylinders, cones and elliptic and hyperbolic paraboloids. A **connected** surface $S$ is a simple surface which cannot be entirely represented by the union of two disjoint open point sets in $\mathbb{R}^3$ where these sets have non-empty intersection with $S$. Hence, for any two arbitrary points, $P_1$ and $P_2$, on $S$ there is a continuous curve which is totally embedded in $S$ with $P_1$ and $P_2$ being its end points. Examples of connected surface are planes, ellipsoids and cylinders.

A **closed** surface is a simple surface with no open edges. Examples of closed surface are spheres and ellipsoids. A **bounded** surface is a surface that can be contained entirely in a sphere of a finite radius such as ellipsoid and torus. A **compact** surface is a surface which is bounded and closed like a torus or a Klein bottle (Fig. 28). If $f$ is a differentiable regular mapping from a surface $S$ to a surface $\bar{S}$, then if $S$ is compact then $\bar{S}$ is compact. If $S_1$ and $S_2$ are two simple surfaces where $S_1$ is connected and $S_2$ is closed and contained in $S_1$, then the two surfaces are equal as point sets. As a result, a simple closed surface cannot be a proper subset of a simple connected surface.

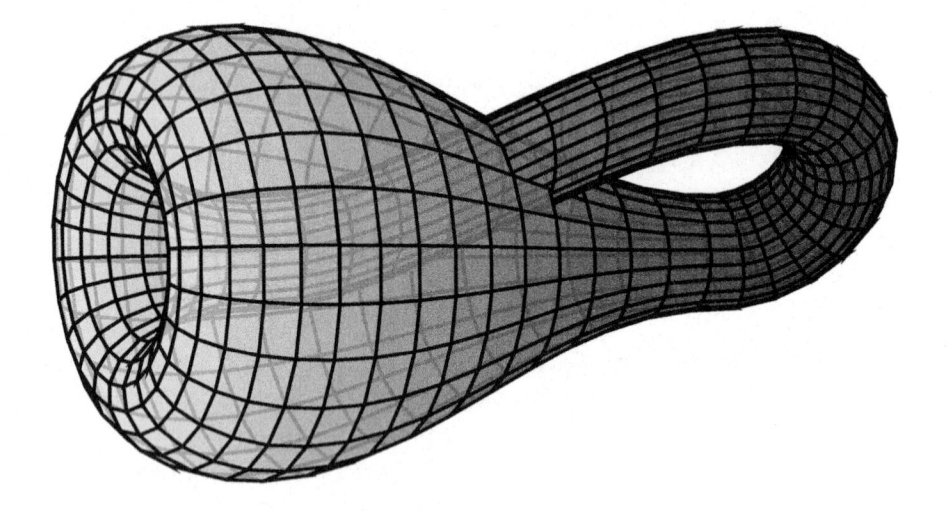

Figure 28: Klein bottle.

An **orientable** surface is a simple surface over which a continuously-varying normal vector can be defined. Hence, spheres, cylinders and tori are orientable surfaces while the Mobius strip (Fig. 1) is a non-orientable surface since a normal vector moved continuously around the strip from a given point will return, following a complete round, to the point in the opposite direction. Similarly, Klein bottle (Fig. 28) is another example of a non-orientable surface. An **oriented** surface is an orientable surface over which the direction of the normal vector is determined. An orientable surface which is connected can be oriented in only one of two possible ways.

An **elementary** surface is a simple surface which possesses a single coordinate patch basis, and hence it is an orientable surface which can be mapped bicontinuously to an open set in the plane. Examples of elementary surface are planes, cones and elliptic

---

[10] There is a more technical and rigorous definition of simple surface which the interested readers are advised to seek in the literature of topology.

## 3.1 General Background about Surfaces

paraboloids. A **developable** surface is a surface that can be flattened into a plane by unfolding without local distortion by compression or stretching. It is called developable because it can be developed into a plane by rolling the surface out on a plane without compression or stretching. A characteristic feature of developable surface is that, like the plane, its Riemann-Christoffel curvature tensor (see § 1.4.10) and Gaussian curvature (see § 4.5) are zero at every point on the surface. Cylinders and cones are obvious examples of developable surface.

A topological property of a surface is a property which is invariant with respect to injective bicontinuous mappings. An example of a topological property is compactness. A differentiable regular mapping from a surface $S$ to a surface $\bar{S}$ is called conformal if it preserves angles between oriented intersecting curves on the surface. The mapping is described as direct if it preserves the sense of the angles and inverse if it reverses it. Technically, the mapping is conformal if there is a function $q(u,v) > 0$ that applies to all patches on the surface such that:

$$a_{\alpha\beta} = q\,\bar{a}_{\alpha\beta} \qquad (\alpha, \beta = 1, 2) \qquad (159)$$

where the unbarred and barred indexed $a$ are the coefficients of the surface covariant metric tensor in $S$ and $\bar{S}$ respectively. The above condition of conformal mapping may be stated by saying that the coefficients of the first fundamental forms of the two surfaces are proportional at their corresponding points. An example of conformal mapping is the stereographic projection (Fig. 29) from the Riemann sphere to a plane. We remark that stereographic projection is a mapping of the unit sphere onto a plane where each point of the sphere is projected, through the line connecting this point to the north pole of the sphere, onto the point of intersection of that line with the plane. This plane is the tangent plane to the sphere at its south pole.

An isometry or isometric mapping of two surfaces is a one-to-one mapping from a surface $S$ to a surface $\bar{S}$ that preserves distances. Hence, any arc in $S$ is mapped onto an arc in $\bar{S}$ with equal length. The two surfaces $S$ and $\bar{S}$ are described as isometric surfaces. An example of isometric mapping is the deformation of a rectangular plane sheet into a cylinder with no local distortion by compression or stretching and hence the two surfaces are isometric since all distances are preserved during this process. Isometry is a symmetric relation and hence the inverse of an isometric mapping is an isometric mapping, that is if $f$ is an isometry from $S$ to $\bar{S}$, then $f^{-1}$ is an isometry from $\bar{S}$ to $S$ (refer to § 6.5 for more details).

An injective mapping from a surface $S$ onto a surface $\bar{S}$ is an isometry *iff* the coefficients of the first fundamental form (see § 3.5) for any patch on $S$ are identical to the corresponding coefficients of the first fundamental form of its image on $\bar{S}$, that is:

$$E = \bar{E} \qquad\qquad F = \bar{F} \qquad\qquad G = \bar{G} \qquad (160)$$

where the unbarred and barred $E, F, G$ are the coefficients of the first fundamental form of the two surfaces at their corresponding points. As seen and will be seen, the coefficients of the first fundamental form are the same as the coefficients of the surface covariant metric

## 3.1 General Background about Surfaces

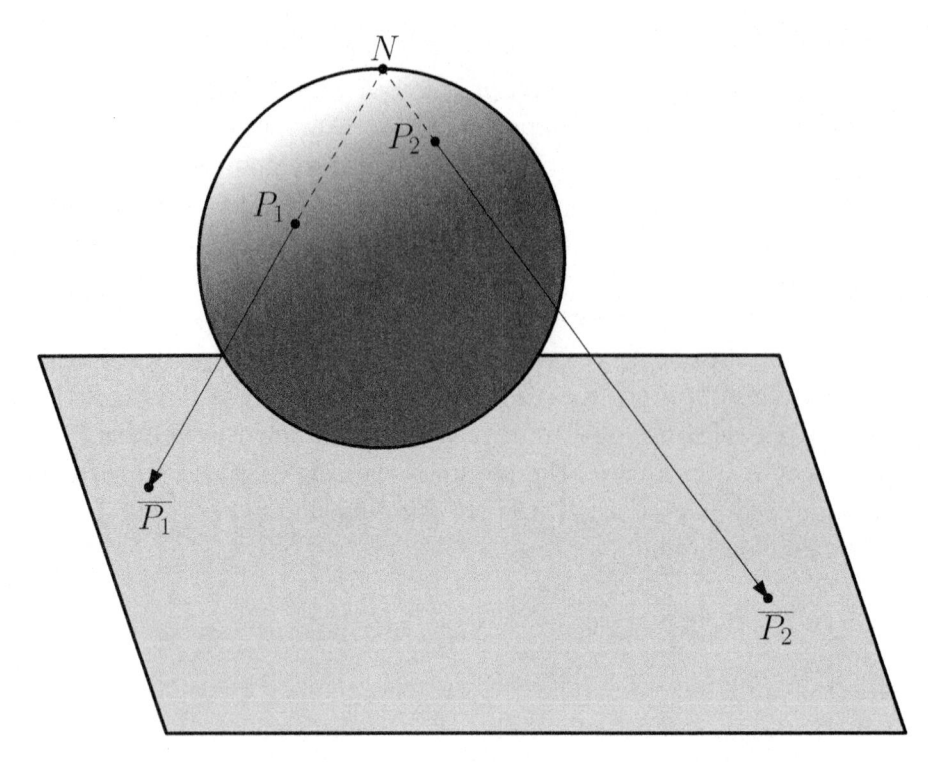

Figure 29: Stereographic projection where the points $P_1$ and $P_2$ on the unit sphere are projected respectively on the points $\overline{P_1}$ and $\overline{P_2}$ on the plane with $N$ representing the north pole of the sphere. The plane is touching the sphere at its south pole.

tensor, that is:

$$a_{11} = E \qquad a_{12} = a_{21} = F \qquad a_{22} = G \qquad (161)$$

and hence these conditions mean that the two surfaces have the same metric at their corresponding points. The mapping that preserves distances but it is not injective is described as local isometry. The statement about the equality of the coefficients of the first fundamental form on the two surfaces also applies to local isometry.

Since intrinsic properties are dependent only on the coefficients of the first fundamental form of the surface, intrinsic properties of the surface are invariant with respect to isometric mappings. As a consequence of the equality of corresponding lengths of two isometric surfaces, the corresponding angles are also equal. However, the reverse is not true, that is a mapping that attains the equality of corresponding angles does not necessarily ensue the equality of corresponding lengths. Hence, isometric mapping is more restrictive than conformal mapping. This means that every isometric mapping is conformal but not every conformal mapping is isometric, so isometric mapping is a subset of conformal mapping. This can be seen by comparing Eqs. 159 and 160 where the latter corresponds to the former with $q = 1$. In fact, conformal mapping can be set up between any two surfaces and in many different ways but this is not always possible for isometric mapping. Isometric mapping also preserves areas of mapped surfaces since it preserves lengths and angles.

A surface generated by the collection of all the tangent lines to a given space curve is called the "tangent surface" of the curve while the tangent lines are called the generators or the rulings of the surface. Similarly, a "branch" of the tangent surface of a curve $C$ at a given point $P$ on the curve refers to the tangent line of $C$ at $P$. In this context, we should remark that the "tangent surface" of a *curve* should not be confused with the aforementioned "tangent plane" of a *surface* at a given point.

The tangent surface of a curve may be demonstrated visually by a taut flexible string connected to the curve where it scans the surface while being directed tangentially at each point of the curve at its base. However, the taut string visualization should extend to both tangential directions to give the full extent of the tangent surface (see § 6.6 for more details).

If $C_e$ is a space curve with a tangent surface $S_T$ and $C_i$ is a curve embedded in $S_T$ and is orthogonal to all the tangent lines of $C_e$ at their intersection points, then $C_i$ is called an involute of $C_e$ while $C_e$ is called an evolute of $C_i$ (see § 5.3).

## 3.2 Mathematical Description of Surfaces

We start by assuming a parametric representation for the surface, where each one of the space coordinates $(x, y, z)$ on the surface is a real differentiable function of the two surface coordinates $(u, v)$. The position vector of a point $P$ on the surface as a function of the surface coordinates is then given by:

$$\mathbf{r}(u, v) = x(u, v)\mathbf{e}_1 + y(u, v)\mathbf{e}_2 + z(u, v)\mathbf{e}_3 \tag{162}$$

where the indexed $\mathbf{e}$ are the Cartesian orthonormal basis vectors in the three directions.

It is also assumed that $\partial_u \mathbf{r}$ and $\partial_v \mathbf{r}$ are linearly independent and hence they are not parallel or anti-parallel, that is:

$$\frac{\partial \mathbf{r}}{\partial u} \times \frac{\partial \mathbf{r}}{\partial v} \neq \mathbf{0} \tag{163}$$

As seen before, this is a sufficient and necessary condition for the surface to be "regular" at a given point. The point is also described as "regular"; otherwise it is "singular" if the condition is violated. The surface is regular on $\Omega$, a closed subset of $\mathbb{R}^2$, if it is regular at each interior point of $\Omega$. The regularity condition guarantees that the surface mapping is one-to-one and possesses a continuous inverse. It also ensures the existence of a tangent plane and a normal vector where this condition is satisfied.

To express the position vector of $P$ in tensor notation, we re-label the space and surface coordinates as:

$$(x, y, z) \equiv (x^1, x^2, x^3) \qquad (u, v) \equiv (u^1, u^2) \tag{164}$$

and hence the position vector of Eq. 162 becomes:

$$\mathbf{r}(u^1, u^2) = x^i(u^1, u^2)\mathbf{e}_i \qquad (i = 1, 2, 3) \tag{165}$$

We note that relabeling the surface coordinates is not necessary in this notation but it will be useful in the future for other tensor notations. To define a surface grid serving

## 3.2 Mathematical Description of Surfaces

as a curvilinear positioning system for the surface, one of the surface coordinate variables is held fixed in turn while the other is varied (Figs. 20 and 30). Hence, each one of the following two surface functions:

$$\mathbf{r}(u^1, c_2) \qquad \mathbf{r}(c_1, u^2) \qquad (166)$$

defines a coordinate curve for the surface, where $c_1$ and $c_2$ are given real constants. These two coordinate curves meet at the common surface point $(c_1, c_2)$. The grid is then generated by varying $c_1$ and $c_2$ uniformly to obtain coordinate curves at regular intervals in its domain.

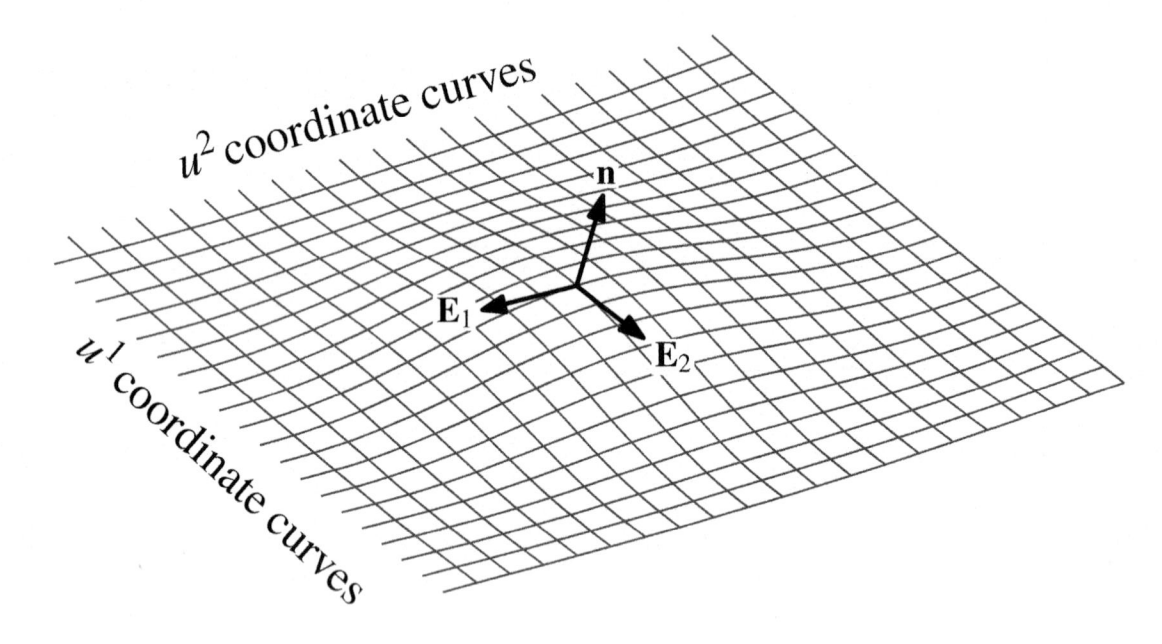

Figure 30: Surface coordinate grid with the surface covariant basis vectors, $\mathbf{E}_1$ and $\mathbf{E}_2$, and the normal vector to the surface $\mathbf{n}$ at a particular point on the surface.

Corresponding to each one of the surface coordinate curves in the above order, a tangent vector to the curve at a given point on the curve is defined by:

$$\mathbf{E}_\alpha = \frac{\partial \mathbf{r}}{\partial u^\alpha} = \frac{\partial x^i}{\partial u^\alpha} \mathbf{e}_i = x^i_\alpha \mathbf{e}_i \qquad (167)$$

where the derivatives are evaluated at that point, and $\alpha = 1, 2$ and $i = 1, 2, 3$. So in brief, $\mathbf{E}_1$ is tangent to the $\mathbf{r}(u^1, c_2)$ coordinate curve and $\mathbf{E}_2$ is tangent to the $\mathbf{r}(c_1, u^2)$ coordinate curve. These tangent vectors serve as a set of basis vectors for the surface, and for each regular point on the surface they generate, by their linear combination, any vector in the surface at that point.[11] They also define, by their linear combination, a plane tangent to the surface at that point. The plane generated by the linear combination of $\mathbf{E}_1$ and $\mathbf{E}_2$

---

[11] This should be understood in an infinitesimal sense or, equivalently, as a vector lying in the tangent plane of the surface at the given point, as will be seen next.

## 3.2 Mathematical Description of Surfaces

is the aforementioned tangent space, $T_P S$, to the surface at point $P$ as described earlier, and hence $\mathbf{E}_1(u_P^1, u_P^2)$ and $\mathbf{E}_2(u_P^1, u_P^2)$ form a basis for this space where the subscript $P$ is a reference to the point $P$.

We should remark that the surface basis vectors, $\mathbf{E}_1$ and $\mathbf{E}_2$, are defined on all points of the surface and not only on the points of the above-described regularly spaced grid of coordinate curves which is presented in that way for pedagogical reasons. Also, a coordinate curve is any surface curve along which only one coordinate variable, $u^1$ or $u^2$, varies regardless of being part of the above grid or not. Another important remark is that the surface coordinate curves are orthogonal at every point on the surface *iff* the surface metric tensor (see § 3.3) is diagonal everywhere on the surface. This is equivalent to having an identically vanishing $F$, which is the coefficient of the first fundamental form, as can be seen for example from Eq. 235 where the dot product will vanish due to the orthogonality of $\mathbf{E}_1$ and $\mathbf{E}_2$ which are the tangents to the coordinate curves. When this condition is satisfied, the coordinate system of the surface is described as orthogonal. Also, the surface coordinate curves are orthogonal at any particular point on the surface *iff* $F = 0$ at that point even if this condition is not satisfied over the entire surface.

Following the definition of the surface basis vectors, $\mathbf{E}_1$ and $\mathbf{E}_2$, a normal vector to the surface at the given point is then defined as the cross product of these tangent basis vectors: $\mathbf{E}_1 \times \mathbf{E}_2$. This normal vector, like $\mathbf{E}_1$ and $\mathbf{E}_2$, is a function of position on the surface and hence in general it varies continuously, in magnitude and direction, as it moves around the surface. This normal vector can be scaled by its magnitude to produce a unit normal vector $\mathbf{n}$ to the surface at that point, that is:[12]

$$\mathbf{n} = \frac{\mathbf{E}_1 \times \mathbf{E}_2}{|\mathbf{E}_1 \times \mathbf{E}_2|} = \frac{\mathbf{E}_1 \times \mathbf{E}_2}{\sqrt{a}} \qquad (168)$$

where $a$ is the determinant of the surface covariant metric tensor (see § 3.3). Based on the cross product rule, the vectors of the triad $\mathbf{E}_1, \mathbf{E}_2, \mathbf{n}$ form a right handed system (see Fig. 30).

On dot producting both sides of Eq. 168 with $\mathbf{n}$, which is a unit vector, we obtain:

$$\mathbf{n} \cdot (\mathbf{E}_1 \times \mathbf{E}_2) = \sqrt{a} \qquad (169)$$

We may also take the modulus of both sides of Eq. 168 (or just compare the denominators of the second equality of Eq. 168) to obtain:

$$|\mathbf{E}_1 \times \mathbf{E}_2| = \sqrt{a} \qquad (170)$$

The surface basis vectors, $\mathbf{E}_\alpha$, are symbolized in full tensor notation by:

$$[\mathbf{E}_\alpha]^i \equiv E_\alpha^i = \frac{\partial x^i}{\partial u^\alpha} = x_\alpha^i \qquad (i = 1, 2, 3 \text{ and } \alpha = 1, 2) \qquad (171)$$

---

[12] Using well-known identities from vector algebra plus what we will see later in this chapter, we obtain:

$$|\mathbf{E}_1 \times \mathbf{E}_2| = \sqrt{|\mathbf{E}_1|^2 |\mathbf{E}_2|^2 - (\mathbf{E}_1 \cdot \mathbf{E}_2)^2} = \sqrt{a_{11} a_{22} - (a_{12})^2} = \sqrt{a}$$

and hence the above equality is fully justified.

## 3.2 Mathematical Description of Surfaces

and hence they can be regarded as 3D contravariant space vectors or as 2D covariant surface vectors (refer to § 3.3 for further details).

It can be shown that the covariant form of the unit normal vector **n** to the surface is given in full tensor notation by:

$$n_i = \frac{1}{2}\underline{\epsilon}^{\alpha\beta}\epsilon_{ijk}x^j_\alpha x^k_\beta \qquad (172)$$

where $x^j_\alpha = \frac{\partial x^j}{\partial u^\alpha}$ and similarly for $x^k_\beta$, and $\underline{\epsilon}^{\alpha\beta}$ and $\epsilon_{ijk}$ are the absolute permutation tensors for the surface and space. The implication of this equation, which defines **n** in terms of the surface basis vectors $x^j_\alpha$ and $x^k_\beta$, is that **n** is a space vector which is independent of the choice of the surface coordinates, $u^1$ and $u^2$, in support of the geometric intuition. Since **n** is normal to the surface, we have:

$$g_{ij}n^i x^j_\alpha = 0 \qquad (173)$$

which is a statement, in tensor notation, that **n** is orthogonal to every vector in the tangent space of the surface at the given point. In this equation, $g_{ij}$ is the space metric tensor.

Although $\mathbf{E}_1$ and $\mathbf{E}_2$ are linearly independent they are not necessarily orthogonal or of unit length. However, they can be orthonormalized as follows:

$$\underline{\mathbf{E}}_1 = \frac{\mathbf{E}_1}{|\mathbf{E}_1|} = \frac{\mathbf{E}_1}{\sqrt{a_{11}}} \qquad \underline{\mathbf{E}}_2 = \frac{a_{11}\mathbf{E}_2 - a_{12}\mathbf{E}_1}{\sqrt{a_{11}a}} \qquad (174)$$

where $a$ is the determinant of the surface covariant metric tensor (see § 3.3), the indexed $a$ are the coefficients of this tensor, and the underlined vectors are orthonormal basis vectors, that is:

$$\underline{\mathbf{E}}_1 \cdot \underline{\mathbf{E}}_1 = 1 \qquad \underline{\mathbf{E}}_2 \cdot \underline{\mathbf{E}}_2 = 1 \qquad \underline{\mathbf{E}}_1 \cdot \underline{\mathbf{E}}_2 = 0 \qquad (175)$$

This can be verified by conducting the dot products of the last equation using the vectors defined in Eq. 174.

The transformation rules from one surface coordinate system to another surface coordinate system, notated with unbarred $(u^1, u^2)$ and barred $(\bar{u}^1, \bar{u}^2)$ symbols respectively, where:

$$u^1 = u^1(\bar{u}^1, \bar{u}^2) \qquad u^2 = u^2(\bar{u}^1, \bar{u}^2) \qquad (176)$$
$$\bar{u}^1 = \bar{u}^1(u^1, u^2) \qquad \bar{u}^2 = \bar{u}^2(u^1, u^2) \qquad (177)$$

are similar to the general rules for the transformation between coordinate systems in a general $n$D space, as explained in the textbooks of tensor analysis.

Following a transformation from an unbarred surface coordinate system to a barred surface coordinate system, the surface becomes a function of the barred coordinates, and a new set of basis vectors for the surface, which are the tangents to the coordinate curves of the barred system, are defined by the following equations:

$$\bar{\mathbf{E}}_1 = \frac{\partial \mathbf{r}}{\partial \bar{u}^1} = \frac{\partial \mathbf{r}}{\partial u^1}\frac{\partial u^1}{\partial \bar{u}^1} + \frac{\partial \mathbf{r}}{\partial u^2}\frac{\partial u^2}{\partial \bar{u}^1} = \mathbf{E}_1\frac{\partial u^1}{\partial \bar{u}^1} + \mathbf{E}_2\frac{\partial u^2}{\partial \bar{u}^1} \qquad (178)$$

## 3.2 Mathematical Description of Surfaces

$$\bar{\mathbf{E}}_2 = \frac{\partial \mathbf{r}}{\partial \bar{u}^2} = \frac{\partial \mathbf{r}}{\partial u^1}\frac{\partial u^1}{\partial \bar{u}^2} + \frac{\partial \mathbf{r}}{\partial u^2}\frac{\partial u^2}{\partial \bar{u}^2} = \mathbf{E}_1 \frac{\partial u^1}{\partial \bar{u}^2} + \mathbf{E}_2 \frac{\partial u^2}{\partial \bar{u}^2} \qquad (179)$$

These equations, which correlate the surface basis vectors in the barred and unbarred surface coordinate systems, can be compactly presented in tensor notation as:

$$\frac{\partial x^i}{\partial \bar{u}^\alpha} = \frac{\partial x^i}{\partial u^\beta}\frac{\partial u^\beta}{\partial \bar{u}^\alpha} \qquad (180)$$

where $i = 1, 2, 3$ and $\alpha, \beta = 1, 2$.

A set of contravariant basis vectors for the surface may also be defined as the gradient of the surface coordinate curves, that is:

$$\mathbf{E}^\alpha = \nabla u^\alpha \qquad (181)$$

In tensor notation, this basis may be given by:

$$[\mathbf{E}^\alpha]_i \equiv E_i^\alpha = \frac{\partial u^\alpha}{\partial x^i} = x_i^\alpha \qquad (182)$$

Hence, these basis vectors can be regarded as 2D contravariant surface vectors or as 3D covariant space vectors.

The contravariant and covariant forms of the surface basis vectors, $\mathbf{E}^\alpha$ and $\mathbf{E}_\alpha$, are obtained from each other by the index shifting operator for the surface, that is:

$$\mathbf{E}_\alpha = a_{\alpha\beta}\mathbf{E}^\beta \qquad \qquad \mathbf{E}^\alpha = a^{\alpha\beta}\mathbf{E}_\beta \qquad (183)$$

where the indexed $a$ are the covariant and contravariant forms of the surface metric tensor (see § 3.3). The contravariant and covariant forms of the surface basis vectors, $\mathbf{E}^\alpha$ and $\mathbf{E}_\alpha$, are reciprocal systems and hence they satisfy the following reciprocity relations:

$$\mathbf{E}_\alpha \cdot \mathbf{E}^\beta = \delta_\alpha^\beta \equiv a_\alpha^\beta \qquad \qquad \mathbf{E}^\alpha \cdot \mathbf{E}_\beta = \delta_\beta^\alpha \equiv a_\beta^\alpha \qquad (184)$$

where the indexed $\delta$ and $a$ are the mixed type of the Kronecker delta and surface metric tensors.

The surface basis vectors in their covariant and contravariant forms, $x_\alpha^i$ and $x_j^\beta$, and the unit normal vector to the surface $n_k$ are linked by the following relation:

$$x_\alpha^i = \underline{\epsilon}^{ijk}\underline{\epsilon}_{\alpha\beta}x_j^\beta n_k \qquad (185)$$

This equation means that the given product (which looks like a vector cross product) of the surface contravariant basis vector $x_j^\beta$ and the unit normal vector $n_k$ produces a surface covariant basis vector $x_\alpha^i$ and hence it is perpendicular to both. Being a surface basis vector implies orthogonality to the unit normal vector, while being a covariant surface basis vector implies here orthogonality to the contravariant surface basis vector.

## 3.3 Surface Metric Tensor

The surface metric tensor is an absolute, rank-2, 2×2 tensor. As we will see, all forms of this tensor (i.e. covariant, contravariant and mixed) are symmetric. In differential geometry, the surface metric tensor $a_{\alpha\beta}$ may be called the first groundform or the fundamental surface tensor. We remark that the coefficients of the metric tensor are real numbers. Following the example of the metric in general $n$D spaces, the surface metric tensor of a 2D surface embedded in a 3D Euclidean flat space with metric $g_{ij} = \delta_{ij}$ is given in its covariant form by:

$$a_{\alpha\beta} = \mathbf{E}_\alpha \cdot \mathbf{E}_\beta = \frac{\partial \mathbf{r}}{\partial u^\alpha} \cdot \frac{\partial \mathbf{r}}{\partial u^\beta} = \frac{\partial x^i}{\partial u^\alpha} \frac{\partial x^i}{\partial u^\beta} \tag{186}$$

where the indexed $x$ and $u$ are the space Cartesian coordinates and the surface coordinates respectively, and $i = 1, 2, 3$ and $\alpha, \beta = 1, 2$.

The surface and space metric tensors in a general Riemannian space with general metric $g_{ij}$ are related by:

$$a_{\alpha\beta} = \mathbf{E}_\alpha \cdot \mathbf{E}_\beta = \frac{\partial \mathbf{r}}{\partial u^\alpha} \cdot \frac{\partial \mathbf{r}}{\partial u^\beta} = g_{ij} \frac{\partial x^i}{\partial u^\alpha} \frac{\partial x^j}{\partial u^\beta} = g_{ij} x^i_\alpha x^j_\beta \tag{187}$$

where $a_{\alpha\beta}$ and $g_{ij}$ are respectively the surface and space covariant metric tensors, the indexed $x$ and $u$ are general coordinates of the space and surface respectively, and $i, j = 1, 2, 3$ and $\alpha, \beta = 1, 2$. It is obvious that Eq. 186 is a special instance of Eq. 187 for the case of a flat space with a Cartesian system where the space metric is the unity tensor.

Eq. 187 is the fundamental relation that provides the crucial link between the surface and its enveloping space. As indicated before, the partial derivatives in this relation, $\frac{\partial x^i}{\partial u^\alpha}$ and $\frac{\partial x^j}{\partial u^\beta}$, may be considered as contravariant rank-1 3D space tensors or as covariant rank-1 2D surface tensors. Hence, a tensor like $\frac{\partial x^i}{\partial u^\alpha}$ is usually labeled as $x^i_\alpha$ to indicate that it represents two surface vectors which are contravariantly-transformed with respect to the three space coordinates $x^i$:

$$x^i_1 = \left( \frac{\partial x^1}{\partial u^1}, \frac{\partial x^2}{\partial u^1}, \frac{\partial x^3}{\partial u^1} \right) \qquad x^i_2 = \left( \frac{\partial x^1}{\partial u^2}, \frac{\partial x^2}{\partial u^2}, \frac{\partial x^3}{\partial u^2} \right) \tag{188}$$

or three space vectors which are covariantly-transformed with respect to the two surface coordinates $u^\alpha$:

$$x^1_\alpha = \left( \frac{\partial x^1}{\partial u^1}, \frac{\partial x^1}{\partial u^2} \right) \qquad x^2_\alpha = \left( \frac{\partial x^2}{\partial u^1}, \frac{\partial x^2}{\partial u^2} \right) \qquad x^3_\alpha = \left( \frac{\partial x^3}{\partial u^1}, \frac{\partial x^3}{\partial u^2} \right) \tag{189}$$

Any surface vector $A^\alpha$ ($\alpha = 1, 2$), defined as a linear combination of the surface basis vectors $\mathbf{E}_1$ and $\mathbf{E}_2$, can also be considered as a space vector $A^i$ ($i = 1, 2, 3$) where the two representations are linked through the relation:

$$A^i = \frac{\partial x^i}{\partial u^\alpha} A^\alpha = x^i_\alpha A^\alpha \qquad (i = 1, 2, 3 \text{ and } \alpha = 1, 2) \tag{190}$$

## 3.3 Surface Metric Tensor

Now, since we have:

$$a_{\alpha\beta} A^\alpha A^\beta = g_{ij} x^i_\alpha x^j_\beta A^\alpha A^\beta \qquad \text{(Eq. 187)} \qquad (191)$$
$$= g_{ij} x^i_\alpha A^\alpha x^j_\beta A^\beta$$
$$= g_{ij} A^i A^j \qquad \text{(Eq. 190)}$$

then we see that the two representations are equivalent, that is they define a vector of the same magnitude and direction.

The contravariant form of the surface metric tensor is defined as the inverse of the surface covariant metric tensor, that is:

$$a^{\alpha\gamma} a_{\gamma\beta} = \delta^\alpha_\beta \qquad\qquad a_{\alpha\gamma} a^{\gamma\beta} = \delta^\beta_\alpha \qquad (192)$$

Since the first fundamental form is positive definite (see § 3.5), and hence $a > 0$, the existence of an inverse is guaranteed. Similar to the metric tensor in general $n$D spaces, the covariant and contravariant forms of the surface metric tensor, $a_{\alpha\beta}$ and $a^{\alpha\beta}$, are used for lowering and raising indices and related tensor operations.

The covariant type of the surface metric tensor $a_{\alpha\beta}$ is given in matrix form by:

$$[a_{\alpha\beta}] = \begin{bmatrix} a_{11} & a_{12} \\ a_{21} & a_{22} \end{bmatrix} = \begin{bmatrix} E & F \\ F & G \end{bmatrix} \qquad (193)$$

where the coefficients $a_{\alpha\beta}$ are as defined above (see Eq. 187), and $E, F, G$ are the coefficients of the first fundamental form (refer to § 3.5). Because the contravariant form of the surface metric tensor $a^{\alpha\beta}$ is the inverse of its covariant form, it is given by:

$$\begin{aligned}[a^{\alpha\beta}] &= \begin{bmatrix} a^{11} & a^{12} \\ a^{21} & a^{22} \end{bmatrix} \\ &= \frac{1}{a_{11}a_{22} - a_{12}a_{21}} \begin{bmatrix} a_{22} & -a_{12} \\ -a_{21} & a_{11} \end{bmatrix} \\ &= \frac{1}{EG - F^2} \begin{bmatrix} G & -F \\ -F & E \end{bmatrix}\end{aligned} \qquad (194)$$

where the symbols are as defined previously. As seen in the above equation, this tensor is symmetric and hence $a^{12} = a^{21}$. As indicated before, the mixed form of the surface metric tensor $a^\alpha_\beta$ is the identity tensor, that is:

$$[a^\alpha_\beta] = [\delta^\alpha_\beta] = \begin{bmatrix} 1 & 0 \\ 0 & 1 \end{bmatrix} \qquad (195)$$

Following the style of the space metric tensor, the surface metric tensor transforms between the barred and unbarred surface coordinate systems as:

$$\bar{a}_{\alpha\beta} = a_{\gamma\delta} \frac{\partial u^\gamma}{\partial \bar{u}^\alpha} \frac{\partial u^\delta}{\partial \bar{u}^\beta} \qquad (196)$$

## 3.3 Surface Metric Tensor

where the indexed $\bar{a}$ and $a$ are the surface covariant metric tensors in the barred and unbarred systems respectively. The contravariant and mixed forms of the surface metric tensor also follow similar transformation rules to their counterparts of the space metric tensor.

Similar to the determinants of the space metric, the determinants of the surface metric in the barred and unbarred coordinate systems are linked through the Jacobian of transformation by the following relation:

$$\bar{a} = J^2 a \qquad (197)$$

where $\bar{a}$ and $a$ are the determinants of the covariant form of the surface metric tensor in the barred and unbarred systems respectively and $J \left(= \left|\frac{\partial u}{\partial \bar{u}}\right|\right)$ is the Jacobian of the transformation between the two surface systems (refer to § 1.4.3). This relation can be obtained directly from Eq. 196 by taking the determinant of the two sides of that equation.

The Christoffel symbols of the first kind $[\alpha\beta, \gamma]$ for the surface are linked to the surface basis vectors and their partial derivatives by the following relation:

$$[\alpha\beta, \gamma] = \frac{\partial \mathbf{E}_\alpha}{\partial u^\beta} \cdot \mathbf{E}_\gamma \qquad (198)$$

Hence, the relation between the partial derivative of the surface covariant metric tensor and the Christoffel symbols of the first kind is given by:

$$\frac{\partial a_{\alpha\beta}}{\partial u^\gamma} = \frac{\partial (\mathbf{E}_\alpha \cdot \mathbf{E}_\beta)}{\partial u^\gamma} = \frac{\partial \mathbf{E}_\alpha}{\partial u^\gamma} \cdot \mathbf{E}_\beta + \mathbf{E}_\alpha \cdot \frac{\partial \mathbf{E}_\beta}{\partial u^\gamma} = [\alpha\gamma, \beta] + [\beta\gamma, \alpha] \qquad (199)$$

A similar relation between the derivative of the surface metric tensor and the Christoffel symbols of the second kind can be obtained from the previous equation by using the index shifting operator of the surface:

$$\frac{\partial a_{\alpha\beta}}{\partial u^\gamma} = a_{\delta\beta}\Gamma^\delta_{\alpha\gamma} + a_{\delta\alpha}\Gamma^\delta_{\beta\gamma} \qquad (200)$$

For a Monge patch of the form $\mathbf{r}(u,v) = (u, v, f(u,v))$, the surface covariant metric tensor $\mathbf{a}$ is given by:

$$\mathbf{a} \equiv \mathbf{I}_S = \begin{bmatrix} 1 + f_u^2 & f_u f_v \\ f_u f_v & 1 + f_v^2 \end{bmatrix} \qquad (201)$$

where the subscripts $u$ and $v$ stand for partial derivatives of $f$ with respect to these surface coordinates, and $\mathbf{I}_S$ is the tensor of the first fundamental form of the surface (see § 3.5). It is noteworthy that scaling a surface up or down by a constant factor $c > 0$, which is equivalent to scaling all the distances on the surface by that factor, can be done by multiplying the surface metric tensor by $c^2$. This can be seen, for example, from Eq. 187.

In the following subsections, we investigate arc length, area and angle between two curves on a surface. All these entities depend in their definition and quantification on the metric tensor. We will see that all these entities, due to their exclusive dependence on the metric tensor, are invariant under isometric transformations. We will also see that their geometric and tensor formulations are identical to those in a general $n$D space with the use of the surface metric tensor and the surface representation of the involved quantities.

### 3.3.1 Arc Length

The length of an infinitesimal line element of a surface curve represents the resultant growth along the element in the $u$ and $v$ directions between its two end points (see Fig. 31). Following the example of the length of an element of arc of a curve embedded in a general $n$D space, the length of an element of arc of a curve on a 2D surface, $ds$, is given in its general form by (see Eqs. 167 and 187):

$$(ds)^2 = d\mathbf{r} \cdot d\mathbf{r} = \frac{\partial \mathbf{r}}{\partial u^\alpha} \cdot \frac{\partial \mathbf{r}}{\partial u^\beta} du^\alpha du^\beta = \mathbf{E}_\alpha \cdot \mathbf{E}_\beta \, du^\alpha du^\beta = a_{\alpha\beta} du^\alpha du^\beta \qquad (202)$$

where $a_{\alpha\beta}$ is the covariant type of the surface metric tensor, $\mathbf{r}$ is the spatial representation of the surface curve and $\alpha, \beta = 1, 2$.

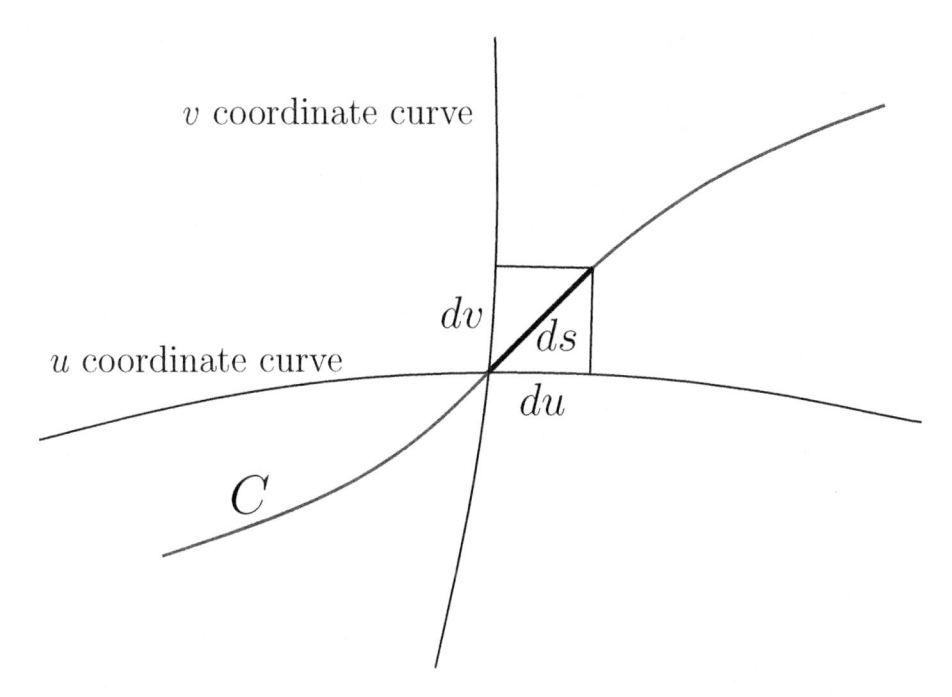

Figure 31: The length of an infinitesimal line element, $ds$, of a surface curve $C$ where $du$ and $dv$ are used to label infinitesimal element growths on the coordinate curves.

From the above formula we have the following identity which is valid at each point of a naturally parameterized surface curve:

$$a_{\alpha\beta} \frac{du^\alpha}{ds} \frac{du^\beta}{ds} = 1 \qquad (203)$$

Based on the above formula (Eq. 202), the length of a segment of a $t$-parameterized curve between a starting point corresponding to $t = t_1$ and a terminal point corresponding to $t = t_2$ is given by:

$$L = \int_I ds \qquad (204)$$

$$= \int_{t_1}^{t_2} \frac{ds}{dt} dt$$

$$= \int_{t_1}^{t_2} \sqrt{a_{\alpha\beta} \frac{du^\alpha}{dt} \frac{du^\beta}{dt}} \, dt$$

$$= \int_{t_1}^{t_2} \sqrt{a_{11} \left(\frac{du^1}{dt}\right)^2 + 2a_{12}\frac{du^1}{dt}\frac{du^2}{dt} + a_{22}\left(\frac{du^2}{dt}\right)^2} \, dt$$

$$= \int_{t_1}^{t_2} \sqrt{E \left(\frac{du^1}{dt}\right)^2 + 2F\frac{du^1}{dt}\frac{du^2}{dt} + G\left(\frac{du^2}{dt}\right)^2} \, dt$$

where $I \subset \mathbb{R}$ is an interval on the real line and $E, F, G$ are the coefficients of the first fundamental form. The fourth equality is based on the equality: $a_{12} = a_{21}$ because the metric tensor is symmetric, as seen earlier.

For a Monge patch of the form $\mathbf{r}(u,v) = (u, v, f(u,v))$, the length of an element of arc of a surface curve is given by:

$$ds = \sqrt{(1 + f_u^2)\, du du + 2 f_u f_v\, du dv + (1 + f_v^2)\, dv dv} \qquad (205)$$

where the subscripts $u$ and $v$ stand for partial derivatives of $f$ with respect to these surface coordinates. The last equation can be obtained directly from Eq. 202 using the coefficients of the metric tensor from Eq. 201.

It is noteworthy that the length of a surface curve is an intrinsic property since it depends on the metric tensor only. Also, the length of a surface curve is invariant with respect to the type of parameterization and isometric transformations.

### 3.3.2 Surface Area

The area of an infinitesimal surface element of a space surface represents the resultant growth in the element as a result of the growth in the $u$ and $v$ directions along the boundary curves that define the element (see Fig. 32).

The area of an infinitesimal element of a surface, $d\sigma$, in the neighborhood of a point $P$ on the surface is given by (see Eqs. 170 and 187):

$$\begin{aligned}
d\sigma &= |d\mathbf{r}_1 \times d\mathbf{r}_2| \qquad (206)\\
&= |\mathbf{E}_1 \times \mathbf{E}_2|\, du^1 du^2 \\
&= \sqrt{|\mathbf{E}_1|^2 |\mathbf{E}_2|^2 - (\mathbf{E}_1 \cdot \mathbf{E}_2)^2}\, du^1 du^2 \\
&= \sqrt{a_{11} a_{22} - (a_{12})^2}\, du^1 du^2 \\
&= \sqrt{EG - F^2}\, du^1 du^2 \\
&= \sqrt{a}\, du^1 du^2
\end{aligned}$$

where $\mathbf{E}_1$ and $\mathbf{E}_2$ are the surface covariant basis vectors, $E, F, G$ are the coefficients of the first fundamental form, $a\ (= a_{11} a_{22} - a_{12} a_{21})$ is the determinant of the surface covariant

### 3.3.2 Surface Area

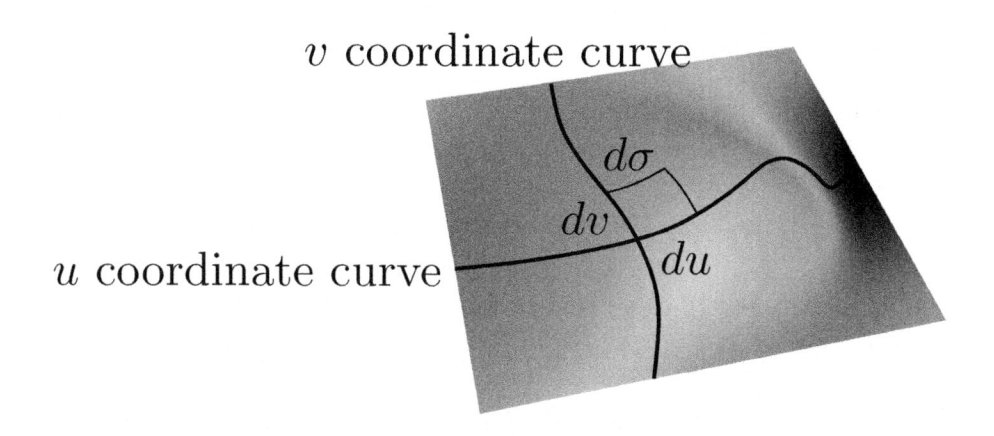

*Figure 32: The area of an infinitesimal surface element, $d\sigma$, of a space surface where $du$ and $dv$ are used to label infinitesimal element growths on the coordinate curves.*

metric tensor and the indexed $a$ are its elements. All the quantities in these expressions belong to the point $P$. We also assume that $du^1 du^2$ is positive.

The area of a surface patch $S: \Omega \to \mathbb{R}^3$, where $\Omega$ is a proper subset of the $\mathbb{R}^2$ parameters plane, is given by:

$$\begin{aligned} \sigma &= \int_\Omega d\sigma \\ &= \iint_\Omega \sqrt{a_{11}a_{22} - (a_{12})^2}\, du^1 du^2 \\ &= \iint_\Omega \sqrt{EG - F^2}\, du^1 du^2 \\ &= \iint_\Omega \sqrt{a}\, du^1 du^2 \end{aligned} \qquad (207)$$

where the symbols are as defined above. The patch $S$ should be injective, sufficiently differentiable, and regular on the interior of $\Omega$. The above formulae for the area are reminder of the volume formulae in a 3D space, so the area can be regraded as a *volume* in a 2D space.

As stated before, the areas of two corresponding surface elements and surface patches on two isometric surfaces are equal. This can be seen from the above formulae since the metric tensor is identical at the corresponding points of two isometric surfaces. For a Monge patch of the form $\mathbf{r}(u,v) = (u, v, f(u,v))$, the surface area is given by:

$$\sigma = \iint_\Omega \sqrt{1 + f_u^2 + f_v^2}\, dudv \qquad (208)$$

where the subscripts $u$ and $v$ stand for partial derivatives of $f$ with respect to these surface coordinates. The last equation can be obtained directly from Eq. 207 using the coefficients of the metric tensor from Eq. 201.

### 3.3.3 Angle Between Two Surface Curves

The angle between two sufficiently smooth surface curves intersecting at a given point on the surface is defined as the angle between their tangent vectors at that point (Fig. 33). As there are two opposite directions for each curve, corresponding to the two senses of traversing the curve, there are two main angles $\theta_1$ and $\theta_2$ such that $\theta_1 + \theta_2 = \pi$. The principal angle between the two curves is usually taken as the smaller of the two angles and hence the directions are determined accordingly. In fact, there are still two possibilities for the directions but this has no significance as far as the angle between the two curves is concerned.

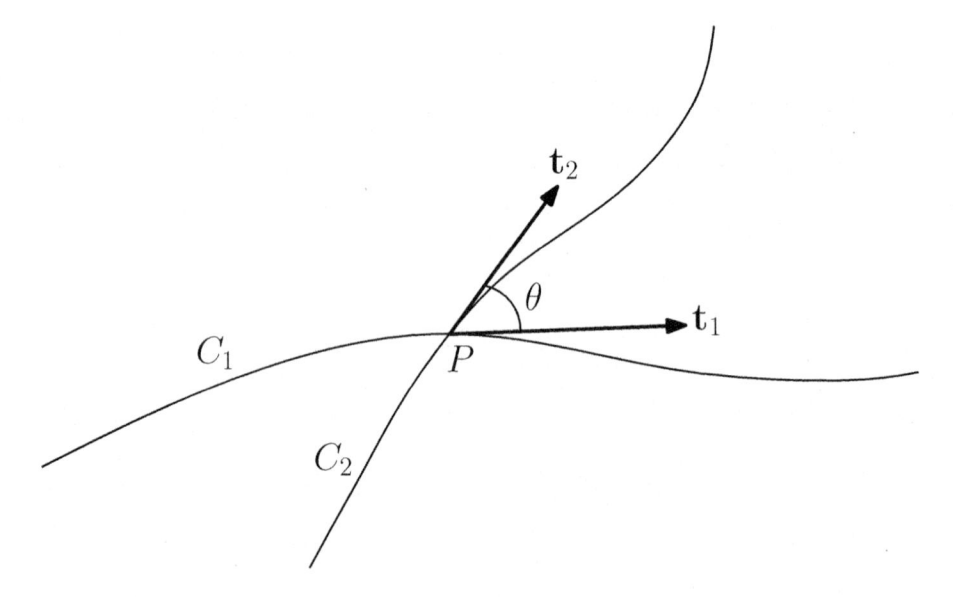

Figure 33: The angle between two surface curves, $C_1$ and $C_2$, as the angle $\theta$ between their tangents, $\mathbf{t}_1$ and $\mathbf{t}_2$, at the point of intersection $P$.

The angle $\theta$ between two surface curves passing through a given point $P$ on the surface with tangent vectors $\mathbf{A}$ and $\mathbf{B}$ at $P$ is given by:

$$\cos \theta = \frac{\mathbf{A} \cdot \mathbf{B}}{|\mathbf{A}||\mathbf{B}|} = \frac{a_{\alpha\beta}A^\alpha B^\beta}{\sqrt{a_{\gamma\delta}A^\gamma A^\delta}\sqrt{a_{\epsilon\zeta}B^\epsilon B^\zeta}} \tag{209}$$

where the indexed $a$ are the elements of the surface covariant metric tensor and the Greek indices run over $1, 2$. If $\mathbf{A}$ and $\mathbf{B}$ are two unit surface vectors with surface representations $A^\alpha$ and $B^\beta$ and space representations $A^i$ and $B^j$ then the angle $\theta$ between the two vectors is given by:

$$\begin{align*}
\cos \theta &= a_{\alpha\beta}A^\alpha B^\beta & \text{(Eq. 209)} \\
&= g_{ij}x^i_\alpha x^j_\beta A^\alpha B^\beta & \text{(Eq. 187)} \\
&= g_{ij}A^i B^j & \text{(Eq. 190)}
\end{align*} \tag{210}$$

where $\alpha, \beta = 1, 2$ and $i, j = 1, 2, 3$. Hence, the surface and space representations of the two vectors define the same angle. The vectors **A** and **B** (whether unit vectors or not) are orthogonal iff $a_{\alpha\beta} A^\alpha B^\beta = 0$ at $P$.

As seen earlier, the coordinate curves at a given point $P$ on a surface are orthogonal iff $a_{12} \equiv F = 0$ at $P$. This can be seen from Eqs. 187 and 235 where the dot products will vanish due to the orthogonality of $\mathbf{E}_1$ and $\mathbf{E}_2$ which are the tangents to the coordinate curves. The corresponding angles of two isometric surfaces, like the corresponding lengths and areas, are equal as can be seen from the above formulae considering that the metric tensor is identical at the corresponding points. However, the reverse is not true in general, that is the equality of angles on two surfaces related by a given mapping, as in the conformal mapping, does not lead to the equality of the corresponding lengths on the two mapped surfaces, as explained before. This is due to the fact that the equality of angles requires the proportionality of the two metric tensors at the corresponding points but not necessarily the equality (see Eq. 159). This may be concluded from Eq. 209 where any proportionality factor will be canceled out.

The sine of the angle $\theta$ between two surface unit vectors, **A** and **B**, is given by:

$$\sin \theta = \underline{\epsilon}_{\alpha\beta} A^\alpha B^\beta \tag{211}$$

where $\underline{\epsilon}_{\alpha\beta}$ is the 2D absolute permutation tensor. The sine in this formula is numerically equal in magnitude to the area of the parallelogram with sides **A** and **B**. Based on the last formula, the sufficient and necessary condition for **A** and **B** to be orthogonal is that:

$$\left| \underline{\epsilon}_{\alpha\beta} A^\alpha B^\beta \right| = 1 \tag{212}$$

## 3.4 Surface Curvature Tensor

The surface curvature tensor is an absolute, rank-2, $2 \times 2$ tensor. As we will see, the covariant and contravariant forms of this tensor are symmetric. The surface curvature tensor $b_{\alpha\beta}$ may also be called the second groundform. We note that the coefficients of the curvature tensor are real numbers. The elements of the surface covariant curvature tensor, $b_{\alpha\beta}$, are given by:

$$b_{\alpha\beta} = -\frac{\partial \mathbf{r}}{\partial u^\alpha} \cdot \frac{\partial \mathbf{n}}{\partial u^\beta} = -\mathbf{E}_\alpha \cdot \frac{\partial \mathbf{n}}{\partial u^\beta} = -\frac{1}{2}\left( \frac{\partial \mathbf{r}}{\partial u^\alpha} \cdot \frac{\partial \mathbf{n}}{\partial u^\beta} + \frac{\partial \mathbf{r}}{\partial u^\beta} \cdot \frac{\partial \mathbf{n}}{\partial u^\alpha} \right) \tag{213}$$

and also by:

$$b_{\alpha\beta} = \frac{\partial^2 \mathbf{r}}{\partial u^\alpha \partial u^\beta} \cdot \mathbf{n} = \frac{\partial \mathbf{E}_\alpha}{\partial u^\beta} \cdot \mathbf{n} = \frac{\partial \mathbf{E}_\alpha}{\partial u^\beta} \cdot \left( \frac{\mathbf{E}_1 \times \mathbf{E}_2}{\sqrt{a}} \right) = \frac{\frac{\partial \mathbf{E}_\alpha}{\partial u^\beta} \cdot (\mathbf{E}_1 \times \mathbf{E}_2)}{\sqrt{a}} \tag{214}$$

where Eq. 168 is used in the last two steps. We note that the equality:

$$\frac{\partial \mathbf{E}_\alpha}{\partial u^\beta} \cdot \mathbf{n} = -\mathbf{E}_\alpha \cdot \frac{\partial \mathbf{n}}{\partial u^\beta} \tag{215}$$

## 3.4 Surface Curvature Tensor

which is seen by comparing Eqs. 213 and 214 is based on the equality:

$$\frac{\partial (\mathbf{E}_\alpha \cdot \mathbf{n})}{\partial u^\beta} = \frac{\partial (0)}{\partial u^\beta} = 0 \qquad (216)$$

plus the product rule of differentiation. Considering Eq. 214, the symmetry of the surface covariant curvature tensor (i.e. $b_{12} = b_{21}$) follows from the fact that:

$$\frac{\partial \mathbf{E}_\alpha}{\partial u^\beta} = \frac{\partial \mathbf{r}}{\partial u^\beta \partial u^\alpha} = \frac{\partial \mathbf{r}}{\partial u^\alpha \partial u^\beta} = \frac{\partial \mathbf{E}_\beta}{\partial u^\alpha} \qquad (217)$$

In full tensor notation, the surface covariant curvature tensor is given by:

$$b_{\alpha\beta} = \frac{1}{2}\epsilon^{\gamma\delta}\epsilon_{ijk}x^i_{\alpha;\beta}x^j_\gamma x^k_\delta = \frac{1}{\sqrt{a}}\epsilon_{ijk}x^i_{\alpha;\beta}x^j_1 x^k_2 \qquad (218)$$

where $a$ is the determinant of the surface covariant metric tensor and the epsilons are the absolute permutation tensors of 2D and 3D spaces in their contravariant and covariant forms. This formula will simplify to the following when the space coordinates are rectangular Cartesian:

$$b_{\alpha\beta} = \frac{1}{\sqrt{a}}\epsilon_{ijk}\frac{\partial^2 x^i}{\partial u^\alpha \partial u^\beta}x^j_1 x^k_2 \qquad (219)$$

The surface curvature tensor obeys the same transformation rules as the surface metric tensor. For example, for the transformation between the barred and unbarred surface coordinate systems we have (see Eq. 196):

$$\bar{b}_{\alpha\beta} = b_{\gamma\delta}\frac{\partial u^\gamma}{\partial \bar{u}^\alpha}\frac{\partial u^\delta}{\partial \bar{u}^\beta} \qquad (220)$$

where the indexed $\bar{b}$ and $b$ are the coefficients of the covariant curvature tensor in the barred and unbarred systems respectively. Similarly, we have (see Eq. 197):

$$\bar{b} = J^2 b \qquad (221)$$

where $\bar{b}$ and $b$ are the determinants of the surface covariant curvature tensor in the barred and unbarred systems respectively and $J \left(= \left|\frac{\partial u}{\partial \bar{u}}\right|\right)$ is the Jacobian of the transformation between the two surface coordinate systems.

The surface covariant curvature tensor is given in matrix form by:

$$\mathbf{b} \equiv [b_{\alpha\beta}] = \begin{bmatrix} b_{11} & b_{12} \\ b_{21} & b_{22} \end{bmatrix} = \begin{bmatrix} e & f \\ f & g \end{bmatrix} \qquad (222)$$

where $e, f, g$ are the coefficients of the second fundamental form of the surface (refer to § 3.6). This is a symmetric matrix as indicated earlier and as seen in the last part of the equation.

The mixed form of the surface curvature tensor $b^\alpha_{\ \beta}$ is given by:

$$[b^\alpha_{\ \beta}] = [a^{\alpha\gamma}b_{\gamma\beta}] = \frac{1}{a}\begin{bmatrix} G & -F \\ -F & E \end{bmatrix}\begin{bmatrix} e & f \\ f & g \end{bmatrix} = \frac{1}{a}\begin{bmatrix} eG - fF & fG - gF \\ fE - eF & gE - fF \end{bmatrix} \qquad (223)$$

## 3.4 Surface Curvature Tensor

where $E, F, G, e, f, g$ are the coefficients of the first and second fundamental forms and $a = EG - F^2$ is the determinant of the surface covariant metric tensor. As seen, the coefficients of $b^\alpha_\beta$ depend on the coefficients of both the first and second fundamental forms. The mixed form $b_\alpha{}^\beta = b_{\alpha\gamma} a^{\gamma\beta}$ is the transpose of the above form.

The contravariant form of the surface curvature tensor is given by:

$$\begin{aligned} \left[b^{\alpha\beta}\right] &= \left[a^{\alpha\gamma} b_\gamma{}^\beta\right] \\ &= \left[b^\alpha{}_\gamma a^{\gamma\beta}\right] \\ &= \frac{1}{a^2} \left[ \begin{array}{cc} eG^2 - 2fFG + gF^2 & fEG - eFG - gEF + fF^2 \\ fEG - eFG - gEF + fF^2 & gE^2 - 2fEF + eF^2 \end{array} \right] \end{aligned} \qquad (224)$$

where the symbols are as explained above. Like the covariant form, the contravariant form is a symmetric tensor as indicated before and as seen above.

As we will see, the trace of the surface mixed curvature tensor $b^\alpha_\beta$ is twice the mean curvature $H$ (see § 4.6), while its determinant is the Gaussian curvature $K$ (see § 4.5), that is:

$$H = \frac{\mathrm{tr}(b^\alpha_\beta)}{2} \qquad K = \det(b^\alpha_\beta) \qquad (225)$$

Because the trace and the determinant of a tensor are its main two invariants under permissible coordinate transformations (refer to similarity transformations of linear algebra), then $H$ and $K$ are invariant, as will be established later in the book.

The surface covariant curvature tensor of a Monge patch of the form $\mathbf{r}(u,v) = (u, v, f(u,v))$ is given by:

$$\mathbf{b} \equiv [b_{\alpha\beta}] = \frac{1}{\sqrt{1 + f_u^2 + f_v^2}} \left[ \begin{array}{cc} f_{uu} & f_{uv} \\ f_{vu} & f_{vv} \end{array} \right] \qquad (226)$$

where the subscripts $u$ and $v$ stand for partial derivatives of $f$ with respect to these surface coordinates. Since $f_{uv} = f_{vu}$, the tensor is symmetric as it should be. The equality: $f_{uv} = f_{vu}$ is based on the well known continuity condition which is fully explained in any standard textbook on calculus.

The covariant curvature tensor of a surface and the Riemann-Christoffel curvature tensor of the first kind are linked by the following relation:

$$R_{\alpha\beta\gamma\delta} = b_{\alpha\gamma} b_{\beta\delta} - b_{\alpha\delta} b_{\beta\gamma} \qquad (227)$$

This relation may also be given in terms of the Riemann-Christoffel curvature tensor of the second kind using the index raising operator for the surface, that is:

$$a^{\alpha\omega} R_{\omega\beta\gamma\delta} = R^\alpha{}_{\beta\gamma\delta} = b^\alpha{}_\gamma b_{\beta\delta} - b^\alpha{}_\delta b_{\beta\gamma} \qquad (228)$$

From Eqs. 88 and 227, it can be concluded that the surface covariant curvature tensor and the surface Christoffel symbols of the first and second kind are related by:

$$b_{\alpha\gamma} b_{\beta\delta} - b_{\alpha\delta} b_{\beta\gamma} = \frac{\partial [\beta\delta, \alpha]}{\partial \gamma} - \frac{\partial [\beta\gamma, \alpha]}{\partial \delta} + [\alpha\delta, \omega] \Gamma^\omega_{\beta\gamma} - [\alpha\gamma, \omega] \Gamma^\omega_{\beta\delta} \qquad (229)$$

Similarly, from Eqs. 89 and 228, it can be concluded that the surface curvature tensor and the surface Christoffel symbols of the second kind are related by:

$$b^\alpha_\gamma b_{\beta\delta} - b^\alpha_\delta b_{\beta\gamma} = \frac{\partial \Gamma^\alpha_{\beta\delta}}{\partial \gamma} - \frac{\partial \Gamma^\alpha_{\beta\gamma}}{\partial \delta} + \Gamma^\omega_{\beta\delta}\Gamma^\alpha_{\omega\gamma} - \Gamma^\omega_{\beta\gamma}\Gamma^\alpha_{\omega\delta} \tag{230}$$

Considering the fact that the 2D Riemann-Christoffel curvature tensor has only one degree of freedom (see § 1.4.10) and hence it possesses a single independent non-vanishing component which is represented by $R_{1212}$, we see that Eq. 227 has only one independent component which is given by:

$$R_{1212} = b_{11}b_{22} - b_{12}b_{21} = b \tag{231}$$

where $b$ is the determinant of the surface covariant curvature tensor.

Equation 227 also shows that each one of the following provisions:

$$R_{\alpha\beta\gamma\delta} = 0 \qquad \text{and} \qquad b_{\alpha\beta} = 0 \qquad (\alpha,\beta,\gamma,\delta=1,2) \tag{232}$$

if satisfied identically is a sufficient and necessary condition for having a flat 2D space. However, for a plane surface all the coefficients of the Riemann-Christoffel curvature tensor and the coefficients of the surface curvature tensor vanish identically throughout the surface, but for a surface which is isometric to plane the coefficients of the Riemann-Christoffel curvature tensor vanish identically but not necessarily the coefficients of the surface curvature tensor. This is based on the fact that having a zero determinant does not imply having a zero tensor (see Eq. 231). So, as stated previously, the provision $R_{\alpha\beta\gamma\delta} = 0$ is a sufficient and necessary condition for having an intrinsically flat space, while the provision $b_{\alpha\beta} = 0$ is a sufficient and necessary condition for having an extrinsically flat space. Having an extrinsically flat space implies having an intrinsically flat space since a curvature that cannot be observed from outside the space cannot be seen from inside the space.

From Eqs. 88 and 227, we see that the Riemann-Christoffel curvature tensor can be expressed in terms of the coefficients of the surface curvature tensor as well as in terms of the coefficients of the surface metric tensor where the two sets of coefficients are linked through Eq. 229. This does not mean that Riemann-Christoffel curvature is an extrinsic property but it means that some intrinsic properties can also be defined in terms of extrinsic parameters. It should be noticed that the sign of the surface curvature tensor (i.e. the sign of its coefficients) is dependent on the choice of the direction of the unit normal vector to the surface, **n**, and hence the sign of the coefficients will be reversed if the direction of **n** is reversed. This can be seen, for example, from Eq. 214.

## 3.5 First Fundamental Form

As indicated previously, the first fundamental form, which is based on the metric, encompasses all the intrinsic information about the surface that a 2D inhabitant of the surface

## 3.5 First Fundamental Form

can obtain from measurements performed on the surface without appealing to an external dimension. In the old books, the first fundamental form may be labeled as the first fundamental quadratic form.

The first fundamental form $I_S$ of the length of an element of arc of a curve on a surface is a quadratic expression given by:

$$\begin{aligned}
I_S &= (ds)^2 \\
&= d\mathbf{r} \cdot d\mathbf{r} \\
&= \frac{\partial \mathbf{r}}{\partial u^\alpha} \cdot \frac{\partial \mathbf{r}}{\partial u^\beta} du^\alpha du^\beta \\
&= \mathbf{E}_\alpha \cdot \mathbf{E}_\beta \, du^\alpha du^\beta \\
&= a_{\alpha\beta} du^\alpha du^\beta \\
&= E(du^1)^2 + 2F \, du^1 du^2 + G(du^2)^2
\end{aligned} \qquad (233)$$

where $E, F, G$, which in general are continuous variable functions of the surface coordinates $u^1$ and $u^2$, are given by:

$$E = a_{11} = \mathbf{E}_1 \cdot \mathbf{E}_1 = \frac{\partial \mathbf{r}}{\partial u^1} \cdot \frac{\partial \mathbf{r}}{\partial u^1} = g_{ij} \frac{\partial x^i}{\partial u^1} \frac{\partial x^j}{\partial u^1} \qquad (234)$$

$$F = a_{12} = \mathbf{E}_1 \cdot \mathbf{E}_2 = \frac{\partial \mathbf{r}}{\partial u^1} \cdot \frac{\partial \mathbf{r}}{\partial u^2} = g_{ij} \frac{\partial x^i}{\partial u^1} \frac{\partial x^j}{\partial u^2} = \mathbf{E}_2 \cdot \mathbf{E}_1 = a_{21} \qquad (235)$$

$$G = a_{22} = \mathbf{E}_2 \cdot \mathbf{E}_2 = \frac{\partial \mathbf{r}}{\partial u^2} \cdot \frac{\partial \mathbf{r}}{\partial u^2} = g_{ij} \frac{\partial x^i}{\partial u^2} \frac{\partial x^j}{\partial u^2} \qquad (236)$$

where the indexed $a$ are the elements of the surface covariant metric tensor, the indexed $x$ are the general coordinates of the enveloping space and $g_{ij}$ is its covariant metric tensor.

For a flat space with a Cartesian coordinate system $x^i$, the space metric is $g_{ij} = \delta_{ij}$ and hence the above equations become:

$$E = \frac{\partial x^i}{\partial u^1} \frac{\partial x^i}{\partial u^1} \qquad (237)$$

$$F = \frac{\partial x^i}{\partial u^1} \frac{\partial x^i}{\partial u^2} \qquad (238)$$

$$G = \frac{\partial x^i}{\partial u^2} \frac{\partial x^i}{\partial u^2} \qquad (239)$$

The first fundamental form can be cast in the following matrix form:

$$\begin{aligned}
I_S &= \begin{bmatrix} du^1 & du^2 \end{bmatrix} \begin{bmatrix} \mathbf{E}_1 \\ \mathbf{E}_2 \end{bmatrix} \cdot \begin{bmatrix} \mathbf{E}_1 & \mathbf{E}_2 \end{bmatrix} \begin{bmatrix} du^1 \\ du^2 \end{bmatrix} \\
&= \begin{bmatrix} du^1 & du^2 \end{bmatrix} \begin{bmatrix} \mathbf{E}_1 \cdot \mathbf{E}_1 & \mathbf{E}_1 \cdot \mathbf{E}_2 \\ \mathbf{E}_2 \cdot \mathbf{E}_1 & \mathbf{E}_2 \cdot \mathbf{E}_2 \end{bmatrix} \begin{bmatrix} du^1 \\ du^2 \end{bmatrix} \\
&= \begin{bmatrix} du^1 & du^2 \end{bmatrix} \begin{bmatrix} E & F \\ F & G \end{bmatrix} \begin{bmatrix} du^1 \\ du^2 \end{bmatrix}
\end{aligned} \qquad (240)$$

## 3.5 First Fundamental Form

$$= \begin{bmatrix} du^1 & du^2 \end{bmatrix} \begin{bmatrix} a_{11} & a_{12} \\ a_{21} & a_{22} \end{bmatrix} \begin{bmatrix} du^1 \\ du^2 \end{bmatrix}$$

$$= \mathbf{v} \, \mathbf{I}_S \, \mathbf{v}^T$$

where $\mathbf{v}$ is a direction vector, $\mathbf{v}^T$ is its transpose, and $\mathbf{I}_S$ is the first fundamental form tensor which is equal to the surface covariant metric tensor. Hence, the matrix associated with the first fundamental form is the covariant metric tensor of the surface. We remark that the notation regarding the dot product operation between two matrices in the first line of Eq. 240 may not be a standard one. We also note that while the coefficients of the first fundamental form depend on the position on the surface and hence they are functions of the surface coordinates but not the direction, the first fundamental form is a function of both position and direction.

The first fundamental form is not a unique characteristic of the surface and hence two geometrically different surfaces as seen from the enveloping space, such as plane and cylinder, can have the same first fundamental form. Such surfaces are different extrinsically as seen from the embedding space although they are identical intrinsically as viewed internally from the surface by a 2D inhabitant.

The first fundamental form is positive definite at regular points of 2D surfaces; hence its coefficients are subject to the following conditions:

$$E > 0 \qquad \text{and} \qquad \det(\mathbf{I}_S) = EG - F^2 > 0 \qquad (241)$$

As indicated earlier, the above conditions imply $G > 0$ since these coefficients are real. It is noteworthy that the condition of positive definiteness may be amended to allow for metrics based on coordinate systems with imaginary coordinates as it is the case in the coordinate systems of the Minkowski space (see § 1.4.7), but this is out of the scope of the present book whose focus is the geometry of space curves and surfaces in a static sense and hence all metrics considered in this book are positive definite.

As indicated above, the first fundamental form encompasses the intrinsic properties of the surface geometry. Hence, as seen in § 3.3, the first fundamental form is used to define and quantify things like arc length, area and angle between curves on a surface based on its qualification as a metric. For example, Eq. 204 shows that the length of a curve segment on a surface is obtained by integrating the square root of the first fundamental form of the surface along the segment.

If a surface $S_1$ can be mapped isometrically (see § 3.1 and § 6.5) onto another surface $S_2$, then the two surfaces have identical first fundamental forms and identical first fundamental form coefficients at their corresponding points, that is:[13]

$$E_1 = E_2 \qquad F_1 = F_2 \qquad G_1 = G_2 \qquad (242)$$

where the subscripts are labels for the two surfaces. Based on Eq. 201, for a Monge patch of the form $\mathbf{r}(u, v) = (u, v, f(u, v))$, the first fundamental form is given by:

$$I_S = \left(1 + f_u^2\right) dudu + 2f_u f_v dudv + \left(1 + f_v^2\right) dvdv \qquad (243)$$

---

[13] In this type of statements, the meaning is that we can find a coordinate system on each surface such that these conditions are satisfied, as indicated early in the book.

where the subscripts $u$ and $v$ stand for partial derivatives of $f$ with respect to these surface coordinates.

## 3.6 Second Fundamental Form

The mathematical entity that characterizes the extrinsic geometry of a surface is the normal vector to the surface. This entity can only be observed externally from outside the surface by an observer in a reference frame in the space that envelops the surface. Hence, the normal vector and all its subsidiaries are strange to a 2D inhabitant to the surface who can only access the intrinsic attributes of the surface as represented by and contained in the first fundamental form. In brief, the 2D inhabitant has no notion of the extra dimension which embraces the normal vector. As a consequence, the variation of the normal vector as it moves around the surface can be used as an indicator to characterize the surface shape from an external point of view and that is how this indicator is exploited in the second fundamental form to represent the extrinsic geometry of the surface as will be seen from the forthcoming formulations of the second fundamental form.

The second fundamental form $II_S$ of a surface, which in the old books may be labeled as the second fundamental quadratic form, is defined by the following quadratic expression:

$$\begin{aligned} II_S &= -d\mathbf{r} \cdot d\mathbf{n} \\ &= -\left(\frac{\partial \mathbf{r}}{\partial u^\alpha} du^\alpha\right) \cdot \left(\frac{\partial \mathbf{n}}{\partial u^\beta} du^\beta\right) \\ &= -\left(\frac{\partial \mathbf{r}}{\partial u^1} du^1 + \frac{\partial \mathbf{r}}{\partial u^2} du^2\right) \cdot \left(\frac{\partial \mathbf{n}}{\partial u^1} du^1 + \frac{\partial \mathbf{n}}{\partial u^2} du^2\right) \\ &= e(du^1)^2 + 2f\, du^1 du^2 + g(du^2)^2 \end{aligned} \qquad (244)$$

where the coefficients of the second fundamental form $e, f, g$ are given by:

$$e = -\frac{\partial \mathbf{r}}{\partial u^1} \cdot \frac{\partial \mathbf{n}}{\partial u^1} = -\mathbf{E}_1 \cdot \frac{\partial \mathbf{n}}{\partial u^1} \qquad (245)$$

$$f = -\frac{1}{2}\left(\frac{\partial \mathbf{r}}{\partial u^1} \cdot \frac{\partial \mathbf{n}}{\partial u^2} + \frac{\partial \mathbf{r}}{\partial u^2} \cdot \frac{\partial \mathbf{n}}{\partial u^1}\right) = -\frac{1}{2}\left(\mathbf{E}_1 \cdot \frac{\partial \mathbf{n}}{\partial u^2} + \mathbf{E}_2 \cdot \frac{\partial \mathbf{n}}{\partial u^1}\right) \qquad (246)$$

$$g = -\frac{\partial \mathbf{r}}{\partial u^2} \cdot \frac{\partial \mathbf{n}}{\partial u^2} = -\mathbf{E}_2 \cdot \frac{\partial \mathbf{n}}{\partial u^2} \qquad (247)$$

In the above equations, $\mathbf{r}(u^1, u^2)$ denotes the spatial representation of the surface, $\mathbf{n}(u^1, u^2)$ is the unit normal vector to the surface and $\alpha, \beta = 1, 2$.

The second fundamental form may also be given by:

$$\begin{aligned} II_S &= d^2\mathbf{r} \cdot \mathbf{n} \\ &= \left(\frac{\partial^2 \mathbf{r}}{\partial u^\alpha \partial u^\beta} du^\alpha du^\beta\right) \cdot \mathbf{n} \\ &= \left(\frac{\partial \mathbf{E}_1}{\partial u^1}(du^1)^2 + 2\frac{\partial \mathbf{E}_1}{\partial u^2} du^1 du^2 + \frac{\partial \mathbf{E}_2}{\partial u^2}(du^2)^2\right) \cdot \mathbf{n} \end{aligned} \qquad (248)$$

## 3.6 Second Fundamental Form

$$= \frac{\partial \mathbf{E}_1}{\partial u^1} \cdot \mathbf{n} \, (du^1)^2 + 2 \frac{\partial \mathbf{E}_1}{\partial u^2} \cdot \mathbf{n} \, du^1 du^2 + \frac{\partial \mathbf{E}_2}{\partial u^2} \cdot \mathbf{n} \, (du^2)^2$$

where $d^2\mathbf{r}$ is the second order differential of the position vector $\mathbf{r}$ of an arbitrary point on the surface in the direction $(du^1, du^2)$. As a result, the coefficients of the second fundamental form can also be given by the following alternative expressions:

$$e = \mathbf{n} \cdot \frac{\partial \mathbf{E}_1}{\partial u^1} = -\frac{\partial \mathbf{n}}{\partial u^1} \cdot \mathbf{E}_1 \tag{249}$$

$$f = \mathbf{n} \cdot \frac{\partial \mathbf{E}_1}{\partial u^2} = \mathbf{n} \cdot \frac{\partial \mathbf{E}_2}{\partial u^1} = -\frac{\partial \mathbf{n}}{\partial u^2} \cdot \mathbf{E}_1 = -\frac{\partial \mathbf{n}}{\partial u^1} \cdot \mathbf{E}_2 \tag{250}$$

$$g = \mathbf{n} \cdot \frac{\partial \mathbf{E}_2}{\partial u^2} = -\frac{\partial \mathbf{n}}{\partial u^2} \cdot \mathbf{E}_2 \tag{251}$$

The two forms of each formula in the last equations are based on the fact that $\mathbf{n} \cdot \mathbf{E}_\alpha = 0$, since $\mathbf{n}$ is perpendicular to the surface basis vectors, plus the product rule of differentiation, that is:

$$\frac{\partial (\mathbf{n} \cdot \mathbf{E}_\alpha)}{\partial u^\beta} = \frac{\partial (0)}{\partial u^\beta} = \mathbf{n} \cdot \frac{\partial \mathbf{E}_\alpha}{\partial u^\beta} + \frac{\partial \mathbf{n}}{\partial u^\beta} \cdot \mathbf{E}_\alpha = 0 \quad \Longrightarrow \quad \mathbf{n} \cdot \frac{\partial \mathbf{E}_\alpha}{\partial u^\beta} = -\frac{\partial \mathbf{n}}{\partial u^\beta} \cdot \mathbf{E}_\alpha \tag{252}$$

The equality: $\mathbf{n} \cdot \partial_2 \mathbf{E}_1 = \mathbf{n} \cdot \partial_1 \mathbf{E}_2$ which is seen in Eq. 250 is based on the equality: $\partial_\alpha \mathbf{E}_\beta = \partial_\beta \mathbf{E}_\alpha$ as stated before (see Eq. 217).

From Eqs. 249-251 plus Eq. 168, we can see that the coefficients of the second fundamental form may also be given by the following expressions:

$$e = \frac{(\mathbf{E}_1 \times \mathbf{E}_2) \cdot \frac{\partial \mathbf{E}_1}{\partial u^1}}{\sqrt{a}} \tag{253}$$

$$f = \frac{(\mathbf{E}_1 \times \mathbf{E}_2) \cdot \frac{\partial \mathbf{E}_1}{\partial u^2}}{\sqrt{a}} \tag{254}$$

$$g = \frac{(\mathbf{E}_1 \times \mathbf{E}_2) \cdot \frac{\partial \mathbf{E}_2}{\partial u^2}}{\sqrt{a}} \tag{255}$$

where $a \,(= a_{11}a_{22} - a_{12}a_{21} = EG - F^2)$ is the determinant of the surface covariant metric tensor. Like the coefficients of the first fundamental form, the coefficients of the second fundamental form are, in general, continuous variable functions of the surface coordinates $u^1$ and $u^2$.

The second fundamental form can also be cast in the following matrix form:

$$\begin{aligned}
II_S &= \begin{bmatrix} du^1 & du^2 \end{bmatrix} \begin{bmatrix} -\mathbf{E}_1 \\ -\mathbf{E}_2 \end{bmatrix} \cdot \begin{bmatrix} \frac{\partial \mathbf{n}}{\partial u^1} & \frac{\partial \mathbf{n}}{\partial u^2} \end{bmatrix} \begin{bmatrix} du^1 \\ du^2 \end{bmatrix} \\
&= \begin{bmatrix} du^1 & du^2 \end{bmatrix} \begin{bmatrix} -\mathbf{E}_1 \cdot \frac{\partial \mathbf{n}}{\partial u^1} & -\mathbf{E}_1 \cdot \frac{\partial \mathbf{n}}{\partial u^2} \\ -\mathbf{E}_2 \cdot \frac{\partial \mathbf{n}}{\partial u^1} & -\mathbf{E}_2 \cdot \frac{\partial \mathbf{n}}{\partial u^2} \end{bmatrix} \begin{bmatrix} du^1 \\ du^2 \end{bmatrix} \\
&= \begin{bmatrix} du^1 & du^2 \end{bmatrix} \begin{bmatrix} \frac{\partial \mathbf{E}_1}{\partial u^1} \cdot \mathbf{n} & \frac{\partial \mathbf{E}_1}{\partial u^2} \cdot \mathbf{n} \\ \frac{\partial \mathbf{E}_2}{\partial u^1} \cdot \mathbf{n} & \frac{\partial \mathbf{E}_2}{\partial u^2} \cdot \mathbf{n} \end{bmatrix} \begin{bmatrix} du^1 \\ du^2 \end{bmatrix}
\end{aligned} \tag{256}$$

## 3.6 Second Fundamental Form

$$= \begin{bmatrix} du^1 & du^2 \end{bmatrix} \begin{bmatrix} e & f \\ f & g \end{bmatrix} \begin{bmatrix} du^1 \\ du^2 \end{bmatrix}$$

$$= \mathbf{v}\,\mathbf{II}_S\,\mathbf{v}^T$$

where $\mathbf{v}$ is a direction vector, $\mathbf{v}^T$ is its transpose, and $\mathbf{II}_S$ is the second fundamental form tensor. As indicated before, the notation regarding the dot product operation may not be standard in the commonly employed matrix notation. Also, although the coefficients of the second fundamental form depend on the position but not the direction, the second fundamental form depends on both.

As seen before, the coefficients of the second fundamental form satisfy the following relations:

$$e = b_{11} \qquad f = b_{12} = b_{21} \qquad g = b_{22} \qquad (257)$$

Hence, the second fundamental form tensor $\mathbf{II}_S$ is the same as the surface covariant curvature tensor $\mathbf{b}$, that is:

$$\mathbf{II}_S = \begin{bmatrix} e & f \\ f & g \end{bmatrix} = \begin{bmatrix} b_{11} & b_{12} \\ b_{21} & b_{22} \end{bmatrix} = \mathbf{b} \qquad (258)$$

As a consequence of the previous statements, we see that the second fundamental form can also be given in a compact tensor notation form by:

$$II_S = b_{\alpha\beta} du^\alpha du^\beta \qquad (259)$$

where the indexed $b$ are the elements of the surface covariant curvature tensor and $\alpha, \beta = 1, 2$.

The second fundamental form of the surface at a given point and in a given direction can also be expressed in terms of the first fundamental form $I_S$ and the normal curvature $\kappa_n$ of the surface at that point and in that direction (see § 4.2) as:

$$II_S = \kappa_n I_S = \kappa_n (ds)^2 \qquad (260)$$

Based on Eqs. 259 and 226, for a Monge patch of the form $\mathbf{r}(u, v) = (u, v, f(u, v))$, the second fundamental form is given by:

$$II_S = \frac{f_{uu} du\,du + 2 f_{uv} du\,dv + f_{vv} dv\,dv}{\sqrt{1 + f_u^2 + f_v^2}} \qquad (261)$$

where the subscripts $u$ and $v$ stand for partial derivatives of $f$ with respect to these surface coordinates.

While the first fundamental form is invariant under isometric transformations, the second fundamental form is invariant under rigid motion transformations. The second fundamental form is also invariant (considering spatially-fixed directions) under permissible surface re-parameterizations that maintain the sense of the normal vector to the surface, $\mathbf{n}$. However, the second fundamental form changes its sign if the sense of $\mathbf{n}$ is reversed. As indicated earlier, while the first fundamental form encompasses the intrinsic geometry of the

## 3.6 Second Fundamental Form

surface, the second fundamental form encompasses its extrinsic geometry. Consequently, the first fundamental form is associated with the surface covariant metric tensor, while the second fundamental form is associated with the surface covariant curvature tensor. A major difference between the two fundamental forms is that while the first fundamental form is positive definite, as stated previously, the second fundamental form is not necessarily positive or definite.

Unlike space curves which are completely defined by specified curvature and torsion, $\kappa$ and $\tau$, providing arbitrary first and second fundamental forms is not a sufficient condition for the existence and determination of a surface with these forms. This is due to the fact that the first and second fundamental forms when independently defined do not provide acceptable determination for the surface unless they satisfy extra compatibility conditions to ensure the consistency of these forms and secure the existence of a surface associated with these forms. In more technical terms, defining six functions $E, F, G, e, f, g$ of class $C^3$ on a subset of $\mathbb{R}^2$ where these functions satisfy the conditions for the coefficients of the first and second fundamental forms (in particular $E, G > 0$ and $EG - F^2 > 0$) does not guarantee the existence of a surface over the given subset with a first fundamental form: $E(du^1)^2 + 2F du^1 du^2 + G(du^2)^2$ and a second fundamental form: $e(du^1)^2 + 2f du^1 du^2 + g(du^2)^2$. Further compatibility conditions relating the first and second fundamental forms are required to fully identify the surface and secure its existence.

The above difference between curves and surfaces may be linked to the fact that the conditions for the curves are established based on the existence theorem for ordinary differential equations where these equations generally have a solution, while the conditions for the surfaces are established based on the existence theorem for partial differential equations which have solutions only when they meet additional integrability conditions. The details should be sought in more extensive books on differential geometry.

Following the statements in the last paragraphs, the required compatibility conditions for the existence of a surface with predefined first and second fundamental forms are given by the Codazzi-Mainardi equations (Eqs. 294-295) plus the following equation:

$$eg - f^2 = F\left[\frac{\partial \Gamma^2_{22}}{\partial u} - \frac{\partial \Gamma^2_{12}}{\partial v} + \Gamma^1_{22}\Gamma^2_{11} - \Gamma^1_{12}\Gamma^2_{12}\right] + \qquad (262)$$
$$E\left[\frac{\partial \Gamma^1_{22}}{\partial u} - \frac{\partial \Gamma^1_{12}}{\partial v} + \Gamma^1_{22}\Gamma^1_{11} + \Gamma^2_{22}\Gamma^1_{12} - \Gamma^1_{12}\Gamma^1_{12} - \Gamma^2_{12}\Gamma^1_{22}\right]$$

where all the symbols are as defined previously and the Christoffel symbols belong to the surface.

From the above statements, the fundamental theorem of space surfaces in differential geometry, which is the equivalent of the fundamental theorem of curves (see § 2.3), emerges. The theorem states that: given six sufficiently smooth functions $E, F, G, e, f, g$ on a subset of $\mathbb{R}^2$ satisfying the following conditions:

1. $E, G > 0$ and $EG - F^2 > 0$, and
2. $E, F, G, e, f, g$ satisfy Eqs. 262 and 294-295,

then there is a unique surface with $E, F, G$ as its first fundamental form coefficients and $e, f, g$ as its second fundamental form coefficients. Hence, if two surfaces meet all these conditions, then they are identical within a rigid motion transformation in space. As a result, the fundamental theorem of surfaces, like the fundamental theorem of curves, provides the existence and uniqueness conditions for surfaces.

On the other hand, according to one of the theorems of Bonnet, if there are two surfaces of class $C^3$, $S_1 : \Omega \to \mathbb{R}^3$ and $S_2 : \Omega \to \mathbb{R}^3$, which are defined over a connected set $\Omega \subseteq \mathbb{R}^2$ and they have identical first and second fundamental forms, then the two surfaces can be mapped on each other by a purely rigid motion transformation. The existence of these surfaces guarantees the compatibility of their first and second fundamental forms and hence they are identical within a rigid motion transformation according to the fundamental theorem of surfaces.

It is noteworthy that two surfaces having identical first fundamental forms but different second fundamental forms may be described as applicable. An example of applicable surfaces is plane and cylinder. Although all applicable surfaces, according to this definition, are isometric since they have identical first fundamental forms, not all isometric surfaces are applicable since two isometric surfaces may also have identical second fundamental forms as it is the case of two planes.

It should be remarked that some authors use $L, M, N$ instead of $e, f, g$ to symbolize the coefficients of the second fundamental form. However, the use of $e, f, g$ is advantageous since they correspond nicely to the coefficients of the first fundamental form $E, F, G$ making the formulae involving the first and second fundamental forms more symmetric and memorable. On the other hand, the use of $L, M, N$ is also advantageous when reciting formulae especially if the formulae contain the coefficients of both fundamental forms; moreover, it is less susceptible to errors in the writing and typing of these formulae.

Another remark is that the coefficient $g$ of the second fundamental form should not be confused with the symbol $g$ of the determinant of the space covariant metric tensor which is commonly used in the literature of tensor calculus and differential geometry but we do not use it in the present book. The coefficient $f$ of the second fundamental form should also be distinguished easily from the symbol $f$ which is widely used in the mathematical literature to symbolize mathematical functions and hence we kept its use in this book for the sake of readability while making some effort to avoid potential confusion.

## 3.6.1 Dupin Indicatrix

On displacing the tangent plane of a surface $S$ at a given point $P$ infinitesimally toward the surface in the orientation of the normal vector of $S$ at $P$, the intersection curves of $S$ with the displaced tangent plane will take a particular and distinctive shape depending on the local shape of $S$ in the neighborhood of $P$ (refer to Fig. 34). This idea is the base for characterizing and quantifying the local shape of a surface at a particular point using the concept of Dupin indicatrix.[14]

---

[14] The purpose of this demonstration, which employs a typical surface with infinitesimal displacement of the tangent plane, is to provide a visual qualitative impression for the idea of Dupin indicatrix. We note that the indicatrix is not necessarily represented by one of these contours on the surface as the indicatrix is a quadratic equation based on the second fundamental form at a particular point $P$ on

Dupin indicatrix at a given point on a sufficiently smooth surface is a function of the coefficients of the second fundamental form at the point and hence it is a function of the surface coordinates. Dupin indicatrix is an indicator of the departure of the surface from the tangent plane in the close proximity of the point of tangency. Accordingly, the second fundamental form coefficients are used in Dupin indicatrix to measure this departure. In quantitative terms, Dupin indicatrix is the family of conic sections given by the following quadratic equation:

$$eu^2 + 2fuv + gv^2 = \pm 1 \qquad (263)$$

where $e, f, g$ are the coefficients of the second fundamental form at the point with the coordinates $u$ and $v$. As seen, Dupin indicatrix, which is represented by an expression similar in form to the second fundamental form, depends on the coefficients of the second fundamental form and the coordinates of the particular point on the surface and hence it is a function of position but, unlike the second fundamental form, it does not depend on the direction.

As a consequence of the previous statements, Dupin indicatrix can be used to classify the surface points with respect to the local shape of the surface as elliptic, parabolic, hyperbolic or flat (see § 4.9). At an elliptic point the Dupin indicatrix is an ellipse or circle,[15] at a parabolic point the Dupin indicatrix becomes two parallel lines, while at a hyperbolic point the Dupin indicatrix becomes two conjugate hyperbolas (see Fig. 35). At a flat point the Dupin indicatrix is not defined. More details about Dupin indicatrix and the local shape of surface will be given in § 4.9. It is noteworthy that in all cases where the Dupin indicatrix is defined, the principal directions (see § 4.4) at the point coincide with the two perpendicular axes of the indicatrix as seen in Fig. 35.

## 3.7 Third Fundamental Form

The third fundamental form $III_S$ of a space surface is defined by:

$$III_S = d\mathbf{n} \cdot d\mathbf{n} = c_{\alpha\beta} du^\alpha du^\beta \qquad (264)$$

where $\mathbf{n}$ is the unit normal vector to the surface at a given point $P$, $c_{\alpha\beta}$ are the coefficients of the third fundamental form at $P$ and $\alpha, \beta = 1, 2$.

The coefficients of the third fundamental form are given in full tensor notation by:

$$c_{\alpha\beta} = g_{ij} n^i_{,\alpha} n^j_{,\beta} \qquad (265)$$

where $g_{ij}$ is the space covariant metric tensor and the indexed $n$ is the unit normal vector to the surface. We note that these coefficients are real numbers. The coefficients of the third fundamental form may also be given by:

$$c_{\alpha\beta} = a^{\gamma\delta} b_{\alpha\gamma} b_{\beta\delta} \qquad (266)$$

where $a^{\gamma\delta}$ is the surface contravariant metric tensor and the indexed $b$ are the coefficients of the surface covariant curvature tensor.

---

the surface, and hence it is not necessarily satisfied by the surface in the extended neighborhood of $P$.
[15] It is circle if the point is umbilical (see § 4.10).

## 3.7 Third Fundamental Form

(a) Elliptic point

(b) Parabolic point

(c) Hyperbolic point

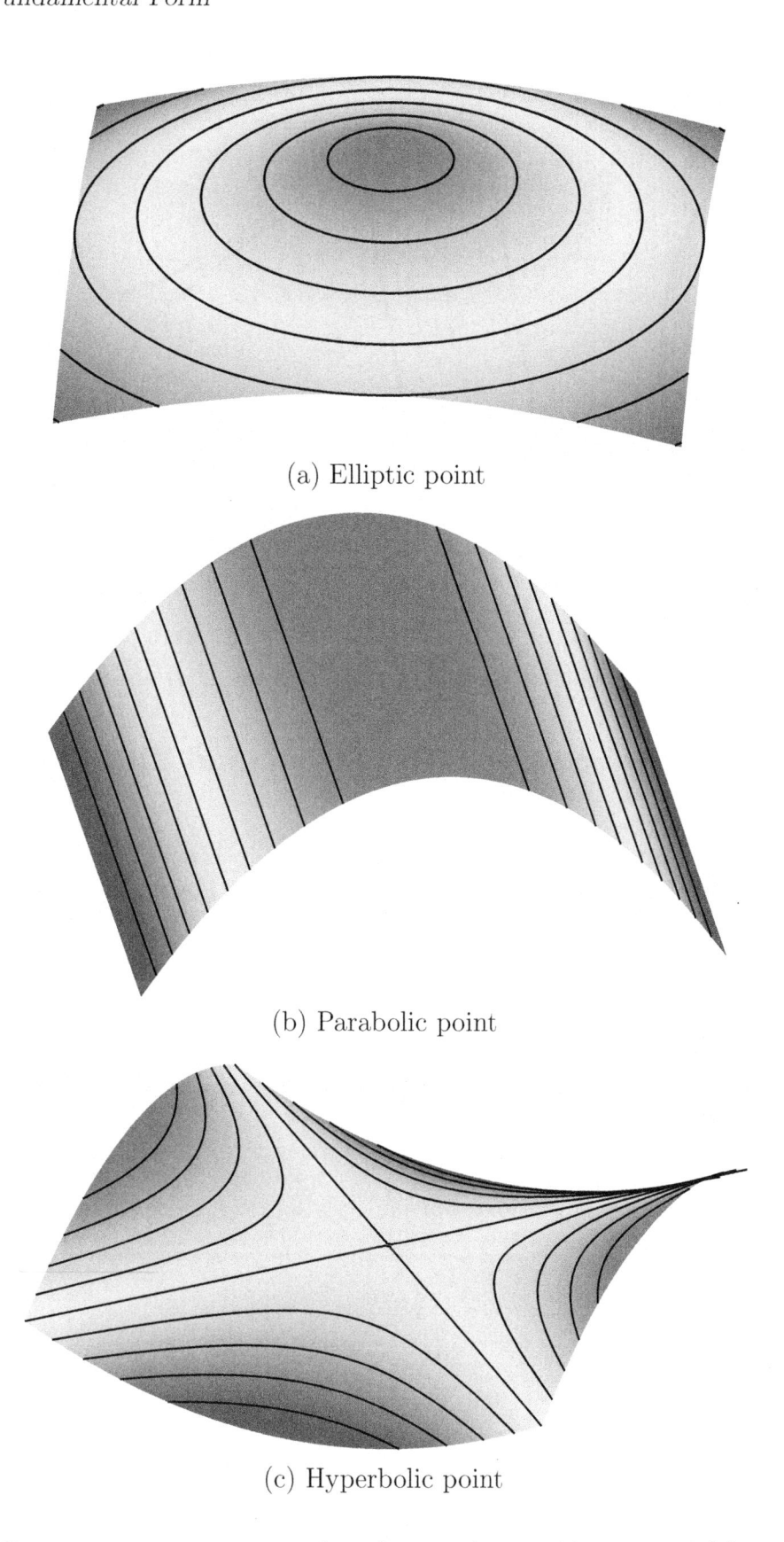

Figure 34: Contour curves on a typical surface in the neighborhood of (a) elliptic point, (b) parabolic point and (c) hyperbolic point.

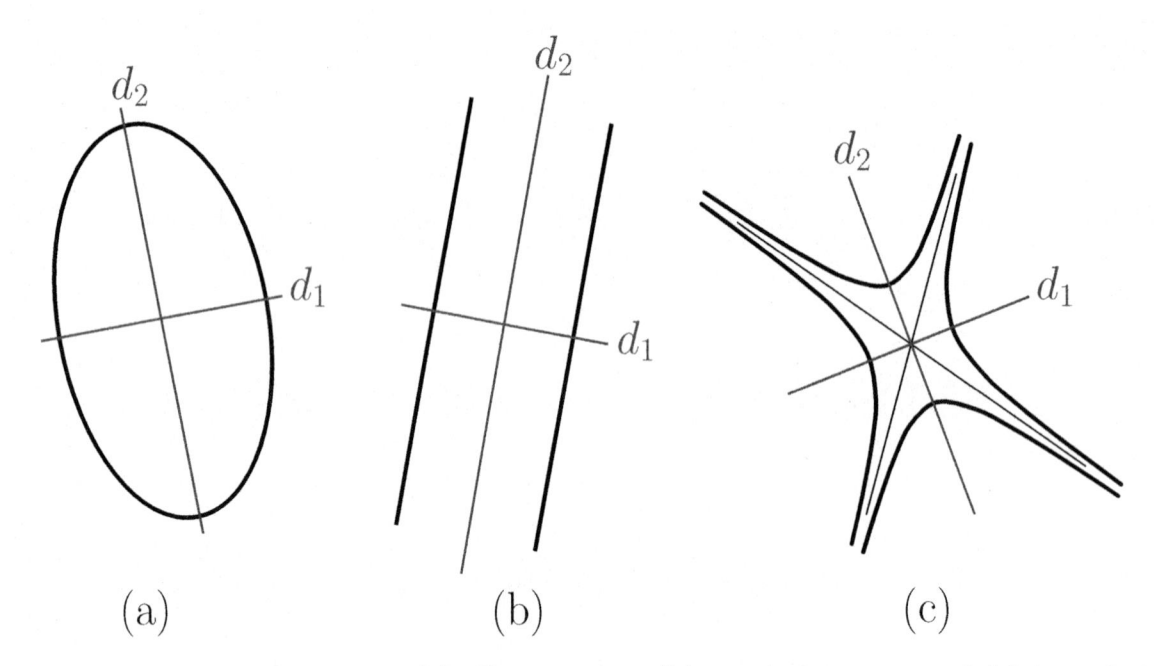

Figure 35: Dupin indicatrix at (a) elliptic point, (b) parabolic point and (c) hyperbolic point. The principal directions (refer to § 4.4) are labeled as $d_1$ and $d_2$.

## 3.8 Relationship between First, Second and Third Fundamental Forms

The first, second and third fundamental forms are linked, through the Gaussian curvature $K$ and the mean curvature $H$ (see § 4.5 and 4.6), by the following relation:

$$K I_S - 2H\, II_S + III_S = 0 \qquad (267)$$

Accordingly, the coefficients of the first, second and third fundamental forms are correlated, through the Gaussian curvature and the mean curvature, by the following relation:

$$K a_{\alpha\beta} - 2H b_{\alpha\beta} + c_{\alpha\beta} = 0 \qquad (268)$$

In fact, Eq. 267 can be obtained from Eq. 268 by multiplying the latter by $du^\alpha du^\beta$ and applying the summation convention. On multiplying both sides of Eq. 268 by $a^{\alpha\beta}$ and shifting the indices we obtain:

$$K a^\alpha_\alpha - 2H b^\alpha_\alpha + c^\alpha_\alpha = 0 \qquad (269)$$

that is:

$$\operatorname{tr}\left(c^\beta_\alpha\right) = 4H^2 - 2K \qquad (270)$$

The transition form Eq. 269 to Eq. 270 is justified by the fact that: $a^\alpha_\alpha = \delta^\alpha_\alpha = \delta^1_1 + \delta^2_2 = 2$ and $H = \frac{b^\alpha_\alpha}{2}$ (see Eq. 383).

## 3.9 Relationship between Surface Basis Vectors and their Derivatives

The focus of this section is the equations of Gauss and Weingarten which, for surfaces, are the analogue of the equations of Frenet-Serret for curves. While the Frenet-Serret formulae express the derivatives of $\mathbf{T}, \mathbf{N}, \mathbf{B}$ as combinations of these vectors using $\kappa$ and $\tau$ as coefficients, the equations of Gauss and Weingarten express the derivatives of $\mathbf{E}_1, \mathbf{E}_2, \mathbf{n}$ as combinations of these vectors with coefficients based on the first and second fundamental forms.

As shown earlier (see § 2.2), three unit vectors can be constructed on each point at which the curvature does not vanish of a class $C^2$ space curve: the tangent $\mathbf{T}$, the principal normal $\mathbf{N}$ and the binormal $\mathbf{B}$. These mutually orthogonal vectors (i.e. $\mathbf{T} \cdot \mathbf{N} = \mathbf{T} \cdot \mathbf{B} = \mathbf{N} \cdot \mathbf{B} = 0$) can serve as a set of basis vectors for the embedding 3D space. Hence, the derivatives of these vectors with respect to the distance traversed along the curve, $s$, can be expressed as combinations of this set, since these derivatives are 3D vectors that reside in this space, as demonstrated by the Frenet-Serret formulae (refer to § 2.5).

Similarly, the surface vectors: $\mathbf{E}_1 = \frac{\partial \mathbf{r}}{\partial u^1}$, $\mathbf{E}_2 = \frac{\partial \mathbf{r}}{\partial u^2}$ and the unit normal vector to the surface, $\mathbf{n}$, at each regular point on a class $C^2$ surface form a basis set for the embedding 3D space and hence their partial derivatives with respect to the surface coordinates, $u^1$ and $u^2$, can be expressed as combinations of this set. The equations of Gauss and Weingarten demonstrate this fact.

The equations of Gauss express the partial derivatives of the surface vectors, $\mathbf{E}_1$ and $\mathbf{E}_2$, with respect to the surface coordinates as combinations of the $\mathbf{E}_1, \mathbf{E}_2, \mathbf{n}$ basis set, that is:

$$\frac{\partial \mathbf{E}_1}{\partial u^1} = \Gamma^1_{11}\mathbf{E}_1 + \Gamma^2_{11}\mathbf{E}_2 + e\mathbf{n} \tag{271}$$

$$\frac{\partial \mathbf{E}_1}{\partial u^2} = \Gamma^1_{12}\mathbf{E}_1 + \Gamma^2_{12}\mathbf{E}_2 + f\mathbf{n} = \frac{\partial \mathbf{E}_2}{\partial u^1} \tag{272}$$

$$\frac{\partial \mathbf{E}_2}{\partial u^2} = \Gamma^1_{22}\mathbf{E}_1 + \Gamma^2_{22}\mathbf{E}_2 + g\mathbf{n} \tag{273}$$

where $e, f, g$ are the coefficients of the second fundamental form. These equations can be expressed compactly, with partial use of tensor notation, as:

$$\frac{\partial \mathbf{E}_\alpha}{\partial u^\beta} = \Gamma^\gamma_{\alpha\beta}\mathbf{E}_\gamma + b_{\alpha\beta}\mathbf{n} \qquad (\alpha, \beta = 1, 2) \tag{274}$$

where the Christoffel symbol of the second kind $\Gamma^\gamma_{\alpha\beta}$ is based on the surface metric, as given by Eqs. 73-78, and $b_{\alpha\beta}$ is the surface covariant curvature tensor. The last equation can be expressed in full tensor notation as:

$$x^i{}_{\alpha,\beta} = \Gamma^\gamma_{\alpha\beta}x^i_\gamma + b_{\alpha\beta}n^i \tag{275}$$

where the symbols and notations are as defined previously. These equations may also be expressed as:

$$x^i{}_{\alpha;\beta} = b_{\alpha\beta}n^i \tag{276}$$

where the covariant derivative notation is in use and a rectangular Cartesian coordinate system for the space is assumed (refer to Eq. 459).

Likewise, the equations of Weingarten express the partial derivatives of the unit normal vector to the surface, $\mathbf{n}$, with respect to the surface coordinates as combinations of the surface vectors, $\mathbf{E}_1$ and $\mathbf{E}_2$, that is:

$$\frac{\partial \mathbf{n}}{\partial u^1} = \frac{fF - eG}{a}\mathbf{E}_1 + \frac{eF - fE}{a}\mathbf{E}_2 \qquad (277)$$

$$\frac{\partial \mathbf{n}}{\partial u^2} = \frac{gF - fG}{a}\mathbf{E}_1 + \frac{fF - gE}{a}\mathbf{E}_2 \qquad (278)$$

where $E, F, G, e, f, g$ are the coefficients of the first and second fundamental forms and $a = EG - F^2$ is the determinant of the surface covariant metric tensor. We note that these expressions are also combinations of the $\mathbf{E}_1, \mathbf{E}_2, \mathbf{n}$ basis set but with vanishing normal components.

The above equations of Weingarten can be expressed compactly, with partial use of tensor notation, as:

$$\frac{\partial \mathbf{n}}{\partial u^\alpha} = -b_\alpha^\beta \mathbf{E}_\beta \qquad (279)$$

where $b_\alpha^\beta$ $(= b_{\alpha\gamma} a^{\gamma\beta})$ is the mixed type of the surface curvature tensor, $b_{\alpha\gamma}$ is the surface covariant curvature tensor and $a^{\gamma\beta}$ is the surface contravariant metric tensor (refer to Eq. 223 and surrounding text). They can also be expressed with full use of tensor notation as:

$$n^i_{,\alpha} = -b_{\alpha\gamma} a^{\gamma\beta} x^i_\beta = -b_\alpha^\beta x^i_\beta \qquad (280)$$

To make sense of this equation, the vector $n^i_{,\alpha}$ is orthogonal to $n^i$ and hence it is parallel to the tangent space of the surface, so it can be expressed as a linear combination of the surface basis vectors $x^i_\beta$, that is: $n^i_{,\alpha} = d_\alpha^\beta x^i_\beta$ for a certain set of coefficients $d_\alpha^\beta = -b_{\alpha\gamma} a^{\gamma\beta}$, as seen above.

The Weingarten equations may also be expressed in matrix form as:

$$\begin{bmatrix} \frac{\partial \mathbf{n}}{\partial u^1} \\ \frac{\partial \mathbf{n}}{\partial u^2} \end{bmatrix} = -\mathbf{II}_S \mathbf{I}_S^{-1} \begin{bmatrix} \mathbf{E}_1 \\ \mathbf{E}_2 \end{bmatrix} \qquad (281)$$

where $\mathbf{II}_S$ is the surface covariant curvature tensor and $\mathbf{I}_S^{-1}$ is the surface contravariant metric tensor. In fact, this is the matrix form of Eq. 279.

The partial derivatives of the unit normal vector to the surface, $\mathbf{n}$, with respect to the surface coordinates, $u^1$ and $u^2$, are linked to the Gaussian curvature $K$ (see § 4.5) and the surface basis vectors, $\mathbf{E}_1$ and $\mathbf{E}_2$, as well as the coefficients of the first and second fundamental forms $E, F, G, e, f, g$ by the following relation:

$$\frac{\partial \mathbf{n}}{\partial u^1} \times \frac{\partial \mathbf{n}}{\partial u^2} = \frac{eg - f^2}{EG - F^2}(\mathbf{E}_1 \times \mathbf{E}_2) = K(\mathbf{E}_1 \times \mathbf{E}_2) \qquad (282)$$

where Eq. 356 is used in the last step.

## 3.9 Relationship between Surface Basis Vectors and their Derivatives

The partial derivatives of the unit normal vector to the surface, $\mathbf{n}$, with respect to the surface coordinates, $u^1$ and $u^2$, are also linked to the Gaussian and mean curvatures, $K$ and $H$, and the coefficients of the first and second fundamental forms $E, F, G, e, f, g$ by the following relations:

$$\frac{\partial \mathbf{n}}{\partial u^1} \cdot \frac{\partial \mathbf{n}}{\partial u^1} = 2eH - EK \tag{283}$$

$$\frac{\partial \mathbf{n}}{\partial u^1} \cdot \frac{\partial \mathbf{n}}{\partial u^2} = 2fH - FK \tag{284}$$

$$\frac{\partial \mathbf{n}}{\partial u^2} \cdot \frac{\partial \mathbf{n}}{\partial u^2} = 2gH - GK \tag{285}$$

The above equations of Weingarten (Eqs. 277-278) can be solved for the surface basis vectors, $\mathbf{E}_1$ and $\mathbf{E}_2$, and hence these vectors can be expressed as combinations of the partial derivatives of the normal vector, $\mathbf{n}$, that is:

$$\mathbf{E}_1 = \frac{fF - gE}{b} \frac{\partial \mathbf{n}}{\partial u^1} + \frac{fE - eF}{b} \frac{\partial \mathbf{n}}{\partial u^2} \tag{286}$$

$$\mathbf{E}_2 = \frac{fG - gF}{b} \frac{\partial \mathbf{n}}{\partial u^1} + \frac{fF - eG}{b} \frac{\partial \mathbf{n}}{\partial u^2} \tag{287}$$

where $b = eg - f^2$ is the determinant of the surface covariant curvature tensor.

From Eq. 274, it can be seen that the coefficients of the surface covariant curvature tensor, $b_{\alpha\beta}$, are the projections of the partial derivative of the surface basis vectors, $\frac{\partial \mathbf{E}_\alpha}{\partial u^\beta}$, in the direction of the unit normal vector to the surface, $\mathbf{n}$, that is:

$$b_{\alpha\beta} = \frac{\partial \mathbf{E}_\alpha}{\partial u^\beta} \cdot \mathbf{n} \tag{288}$$

This can also be seen directly from the definition of the coefficients of the second fundamental form, i.e. Eqs. 249-251.

As indicated above, the essence of the above equations of Gauss and Weingarten is that the partial derivatives of $\mathbf{E}_1$, $\mathbf{E}_2$ and $\mathbf{n}$ can be represented as combinations of these vectors with coefficients obtained from the coefficients of the first and second fundamental forms and their partial derivatives. The above equations are various demonstrations of this fact.

For a Monge patch of the form $\mathbf{r}(u,v) = (u, v, f(u,v))$, the Gauss equations are given by:

$$\frac{\partial \mathbf{E}_1}{\partial u} = \frac{1}{1 + f_u^2 + f_v^2} \left( f_u f_{uu} \mathbf{E}_1 + f_v f_{uu} \mathbf{E}_2 + f_{uu} \sqrt{1 + f_u^2 + f_v^2}\, \mathbf{n} \right) \tag{289}$$

$$\frac{\partial \mathbf{E}_1}{\partial v} = \frac{1}{1 + f_u^2 + f_v^2} \left( f_u f_{uv} \mathbf{E}_1 + f_v f_{uv} \mathbf{E}_2 + f_{uv} \sqrt{1 + f_u^2 + f_v^2}\, \mathbf{n} \right) = \frac{\partial \mathbf{E}_2}{\partial u} \tag{290}$$

$$\frac{\partial \mathbf{E}_2}{\partial v} = \frac{1}{1 + f_u^2 + f_v^2} \left( f_u f_{vv} \mathbf{E}_1 + f_v f_{vv} \mathbf{E}_2 + f_{vv} \sqrt{1 + f_u^2 + f_v^2}\, \mathbf{n} \right) \tag{291}$$

where the subscripts $u$ and $v$ represent partial derivatives of $f$ with respect to the surface coordinates $u$ and $v$.

### 3.9.1 Codazzi-Mainardi Equations

Similarly, for a Monge patch of the form $\mathbf{r}(u,v) = (u, v, f(u,v))$, the Weingarten equations are given by:

$$\frac{\partial \mathbf{n}}{\partial u} = \frac{(f_u f_v f_{uv} - f_{uu} f_v^2 - f_{uu})\mathbf{E}_1 + (f_u f_v f_{uu} - f_u^2 f_{uv} - f_{uv})\mathbf{E}_2}{\sqrt{(1 + f_u^2 + f_v^2)^3}} \qquad (292)$$

$$\frac{\partial \mathbf{n}}{\partial v} = \frac{(f_u f_v f_{vv} - f_{uv} f_v^2 - f_{uv})\mathbf{E}_1 + (f_u f_v f_{uv} - f_u^2 f_{vv} - f_{vv})\mathbf{E}_2}{\sqrt{(1 + f_u^2 + f_v^2)^3}} \qquad (293)$$

### 3.9.1 Codazzi-Mainardi Equations

From the aforementioned equations of Gauss and Weingarten, supported by further compatibility conditions, the following equations, called Codazzi or Codazzi-Mainardi equations, can be derived:

$$\frac{\partial b_{12}}{\partial u^1} - \frac{\partial b_{11}}{\partial u^2} = b_{22}\Gamma_{11}^2 - b_{12}\left(\Gamma_{12}^2 - \Gamma_{11}^1\right) - b_{11}\Gamma_{12}^1 \qquad (294)$$

$$\frac{\partial b_{22}}{\partial u^1} - \frac{\partial b_{21}}{\partial u^2} = b_{22}\Gamma_{12}^2 - b_{12}\left(\Gamma_{22}^2 - \Gamma_{12}^1\right) - b_{11}\Gamma_{22}^1 \qquad (295)$$

where the Christoffel symbols are based on the surface metric. These equations can be expressed compactly in tensor notation as:

$$\frac{\partial b_{\alpha\beta}}{\partial u^\gamma} - \frac{\partial b_{\alpha\gamma}}{\partial u^\beta} = b_{\delta\beta}\Gamma_{\alpha\gamma}^\delta - b_{\delta\gamma}\Gamma_{\alpha\beta}^\delta \qquad (296)$$

Now, if we arrange the terms of the last equation and subtract the term $b_{\alpha\delta}\Gamma_{\gamma\beta}^\delta$ from both sides we obtain:

$$\frac{\partial b_{\alpha\beta}}{\partial u^\gamma} - b_{\delta\beta}\Gamma_{\alpha\gamma}^\delta - b_{\alpha\delta}\Gamma_{\gamma\beta}^\delta = \frac{\partial b_{\alpha\gamma}}{\partial u^\beta} - b_{\delta\gamma}\Gamma_{\alpha\beta}^\delta - b_{\alpha\delta}\Gamma_{\gamma\beta}^\delta \qquad (297)$$

which can be expressed compactly, using the covariant derivative notation (see § 7), as:

$$b_{\alpha\beta;\gamma} = b_{\alpha\gamma;\beta} \qquad (298)$$

The Codazzi-Mainardi equations in the form given by Eq. 298 reveal that there are only two independent components for these equations because, adding to the fact that all the indices range over 1 and 2 and hence we have only eight components, the covariant derivative according to Eq. 298 is symmetric in its last two indices (i.e. $\beta$ and $\gamma$), and the covariant curvature tensor is symmetric in its two indices (i.e. $b_{\alpha\beta} = b_{\beta\alpha}$). These two independent components are given by:

$$b_{\alpha\alpha;\beta} = b_{\alpha\beta;\alpha} \qquad (299)$$

where $\alpha \neq \beta$ and there is no summation over $\alpha$. On writing these equations in full, using the covariant derivative expression (i.e. $b_{\alpha\beta;\gamma} = \frac{\partial b_{\alpha\beta}}{\partial u^\gamma} - b_{\delta\beta}\Gamma_{\alpha\gamma}^\delta - b_{\alpha\delta}\Gamma_{\beta\gamma}^\delta$) and noting that one term of the covariant derivative expression is the same on both sides and hence it drops away, we have:

$$\frac{\partial b_{\alpha\alpha}}{\partial u^\beta} - b_{\alpha\delta}\Gamma^\delta_{\alpha\beta} = \frac{\partial b_{\alpha\beta}}{\partial u^\alpha} - b_{\delta\beta}\Gamma^\delta_{\alpha\alpha} \qquad (\alpha \neq \beta, \text{ no sum on } \alpha) \tag{300}$$

It should be remarked that as a consequence of the aforementioned two symmetries, the covariant derivative of the surface covariant curvature tensor, $b_{\alpha\beta;\gamma}$, is fully symmetric in all of its indices.

There is also another more general equation called (according to some authors) the Gauss-Codazzi equation which is given by:

$$R^\delta_{\alpha\beta\gamma} x^i_\delta = x^i_\delta b^\delta_\beta b_{\alpha\gamma} - x^i_\delta b^\delta_\gamma b_{\alpha\beta} + n^i b_{\alpha\beta;\gamma} - n^i b_{\alpha\gamma;\beta} \tag{301}$$

The tangential component of this equation represents *Theorema Egregium* (in the form given by Eq. 228) while its normal component represents the Codazzi equation (in the form given by Eq. 298). The reader is advised to refer to § 4.7 about the essence of *Theorema Egregium* as an expression of the fact that certain types of curvature are intrinsic properties to the surface and hence they can be expressed in terms of purely intrinsic parameters obtained from the first fundamental form.

## 3.10 Sphere Mapping

Sphere mapping or Gauss mapping is a correlation between the points of a surface and the unit sphere where each point on the surface is projected onto its unit normal as a point on the unit sphere which is centered at the origin of coordinates. This sort of mapping for surfaces is similar to the spherical indicatrix mapping (see § 5.5) for space curves. In technical terms, let $S$ be a surface embedded in an $\mathbb{R}^3$ space and $S_1$ represents the origin-centered unit sphere in this space, then Gauss mapping is given by:

$$\{N : S \to S_1,\ N(P) = \check{P}\} \tag{302}$$

where the point $P(x, y, z)$ on the trace of $S$ is mapped by $N$ onto the point $\check{P}(\check{x}, \check{y}, \check{z})$ on the trace of the unit sphere with $x, y, z$ being the coordinates of $P$ and $\check{x}, \check{y}, \check{z}$ being the coordinates of the origin-based position vector of the normal vector to the surface, $\mathbf{n}$, at $P$. To have a single-valued sphere mapping, the functional relation representing the surface $S$ should be one-to-one.

The image $\bar{\mathfrak{S}}$ on the unit sphere of a Gauss mapping of a patch $\mathfrak{S}$ on a surface $S$ is called the spherical image of $\mathfrak{S}$. The limit of the ratio of the area of a region $\bar{\mathfrak{R}}$ on the spherical image to the area of the corresponding region $\mathfrak{R}$ on the surface $S$ in the neighborhood of a given point $P$ on $S$ equals the absolute value of the Gaussian curvature $|K|$ at $P$ as $\mathfrak{R}$ shrinks to the point $P$, that is:

$$\lim_{\mathfrak{R} \to P} \frac{\sigma(\bar{\mathfrak{R}})}{\sigma(\mathfrak{R})} = |K_P| \tag{303}$$

where $\sigma$ stands for area, and $K_P$ is the Gaussian curvature at $P$. The tendency of $\mathfrak{R}$ to $P$ should be understood in the given sense.

At a given point $P$ on a surface, where the Gaussian curvature is non-zero, there exists a neighborhood $\mathcal{N}$ of $P$ where an injective mapping can be established between $\mathcal{N}$ and its spherical image $\bar{\mathcal{N}}$. A conformal correspondence can be established between a surface and its spherical image *iff* the surface is a sphere or a minimal surface (see § 6.7). We should remark that although the term "spherical image" is related to surfaces in the context of sphere mapping, it can also be used for curves since curves can also have spherical images in a well defined sense.

## 3.11 Global Surface Theorems

In this small section we state, within the following bullet points, a few global theorems related to surfaces to have a taste of this field of differential geometry of surfaces whose investigation is not the main objective of the present book.

- Planes are the only connected surfaces of class $C^2$ whose all points are flat.
- Spheres are the only connected closed surfaces of class $C^3$ whose all points are spherical umbilical (see § 4.10).
- Spheres are the only connected compact surfaces of class $C^3$ with constant Gaussian curvature.
- Spheres are the only connected compact surfaces with constant mean curvature and positive Gaussian curvature.
- The tangent plane to a cylinder or a cone is constant along their generators.
- Any compact surface in $\mathbb{R}^3$ should have points with positive Gaussian curvature.
- Any compact surface in $\mathbb{R}^3$, excluding sphere, should have points with negative Gaussian curvature.
- The reader is also referred to § 4.8 for the global form of the Gauss-Bonnet theorem.

## 3.12 Exercises

3.1 Give the mathematical definition of space surface and explain the difference between a surface and its trace according to this definition.

3.2 What is the mathematical condition for a surface to be regular at a particular point in terms of its basis vectors?

3.3 State the three main mathematical methods for defining a space surface and compare them explaining any advantages or disadvantages in using one of these methods or the others in various contexts.

3.4 Classify the three methods of the last question into two main categories and discuss these categories (see § 1.4.3).

3.5 What "coordinate patch of class $C^n$" means? What are the mathematical conditions that should be satisfied by such a patch?

3.6 Show that a Monge patch of the form $\mathbf{r}(u,v) = (u, v, f(u,v))$ is regular of class $C^n$ if $f$ is of this class.

## 3.12 Exercises

3.7 Give a rigorous definition of "tangent vector" of a surface curve at a particular point on the surface.

3.8 How the tangent plane of a surface at a particular point is related to the basis vectors of the surface at that point?

3.9 Find the equation of the tangent plane to the ellipsoid which is represented parametrically by: $\mathbf{r}(\theta, \phi) = (2\sin\theta \cos\phi, 1.5\sin\theta \sin\phi, 0.5\cos\theta)$ at the point with $\theta = 1.3$ and $\phi = 0.72$.

3.10 Find the equation of the tangent plane of a surface represented parametrically by: $\mathbf{r}(u, v) = (6v, 2u, 1.4u^2 + 6)$ at the point with $u = 1.2$ and $v = 3.6$.

3.11 Show that for a circular cone the tangent plane at all points of any one of its generators is the same.

3.12 Discuss the following statement explaining its meaning in simple words: "The tangent space at a specific point $P$ of a surface is a property of the surface at $P$ and hence it is independent of the patch that contains $P$".

3.13 Does the tangent space of a surface at a given point depend on the particular parameterization of the surface?

3.14 Find the equation of the plane passing through the point $(-1, 3, -9)$ and spanned by the two vectors $(3, 0.5, 1.2)$ and $(0.9, 3, 6.8)$.

3.15 Define, mathematically, the normal unit vector $\mathbf{n}$ of a surface at a given point in terms of the two basis vectors of the surface at that point.

3.16 Calculate, symbolically, the normal unit vector $\mathbf{n}$ at a general point of a surface defined by: $\mathbf{S}(x, y) = (x, y, f)$ where $f = f(x, y)$ is a differentiable function.

3.17 Find the equation of the normal line of a hyperboloid of one sheet given by: $\mathbf{r}(\xi, \theta) = (1.6\cosh\xi \cos\theta, 2.1\cosh\xi \sin\theta, 0.4\sinh\xi)$ at the point with $\xi = 3.2$ and $\theta = 1.5$.

3.18 Find the equation of the normal line of a surface represented by: $\mathbf{r}(u, v) = (3u, u^2 + v, 5v)$ at the point with $u = 2.5$ and $v = -1.8$.

3.19 Define "Monge patch" giving its three forms. Which of these forms is the most common in use?

3.20 Define, briefly, the following terms with some examples representing these concepts: simple surface, simply connected region on a surface, closed surface, compact surface, elementary surface, oriented surface and developable surface.

3.21 Which of the following is a simple surface and which is not: cylinder, hyperboloid of two sheets, torus, Klein bottle, and elliptic paraboloid?

3.22 Is the surface represented by the equation: $x^2 + y^2 + z^2 = 9$ compact? What about the surface represented by: $x^2 - y^2 - z^2 = 4$?

3.23 Why the Mobius strip is not an orientable surface? Give another example of a non-orientable surface.

3.24 What is conformal mapping? What is direct and inverse conformal mapping?

3.25 Describe stereographic mapping making a simple sketch representing this type of mapping.

3.26 Prove that stereographic mapping is conformal.

3.27 Define isometric mapping giving an example of such mapping between two types of surface.

## 3.12 Exercises

3.28 What are the mathematical conditions for two surfaces to be isometric? Does isometry relate to the intrinsic or extrinsic properties of the surface and why?

3.29 What is the distinctive property of an isometric relation between two surfaces in terms of angles, arc lengths and areas defined on the two surfaces?

3.30 What is the relation between conformal mapping and isometric mapping?

3.31 What is local isometry? What is the difference between local isometry and global isometry?

3.32 Show that Eq. 160 applies to local isometric mapping.

3.33 A surface $S_1$ is mapped isometrically onto another surface $S_2$. How the intrinsic properties of $S_1$ will be affected by this mapping?

3.34 Make a clear distinction between the tangent surface of a curve and the tangent plane of a surface describing each of these briefly.

3.35 What is the meaning of "branch of the tangent surface of a curve $C$ at a given point $P$ on the curve"?

3.36 Give a brief definition of involute and evolute.

3.37 Write down a general mathematical relation representing the position vector of a point $P$ on a space surface as a function of the surface coordinates, $u^1$ and $u^2$, where the surface is embedded in a 3D Euclidean space coordinated by a rectangular Cartesian system.

3.38 Describe, in detail, how a coordinate grid is constructed on a space surface with a clear definition of the coordinate curves used to build this grid.

3.39 Make a simple and fully labeled sketch of a space surface coordinated by a curvilinear grid with the two covariant surface basis vectors and the unit normal vector to the surface at one point.

3.40 A surface is represented spatially by: $\mathbf{r}(u,v) = (u, 5, 2v)$. Discuss the type and the main properties of this surface.

3.41 Define, symbolically, the covariant basis vectors of a space surface in terms of the coordinates of the ambient space $x^i$ and the coordinates of the surface $u^\alpha$.

3.42 Give the symbol used in tensor notation to represent the surface basis vectors $\mathbf{E}_\alpha$. From analyzing this symbol, describe how the surface basis vectors can be regarded as covariant and contravariant vectors at the same time.

3.43 Write, in full tensor notation, the equation representing the covariant form of the unit normal vector to a space surface.

3.44 Although $\mathbf{E}_1$ and $\mathbf{E}_2$ are linearly independent at the regular points of the surface they are not necessarily orthogonal or of unit length. Is it possible to construct an orthonormal set of basis vectors from $\mathbf{E}_1$ and $\mathbf{E}_2$? If so, how?

3.45 Following a transformation from an unbarred surface coordinate system to a barred surface coordinate system, what are the mathematical expressions representing the barred basis set, $\bar{\mathbf{E}}_1$ and $\bar{\mathbf{E}}_2$, in terms of the unbarred basis set, $\mathbf{E}_1$ and $\mathbf{E}_2$? Give these expressions in vector and tensor notations.

3.46 Define, descriptively and mathematically, the set of contravariant basis vectors for a space surface discussing how these vectors can be regarded as covariant and contravariant vectors simultaneously.

## 3.12 Exercises

3.47 How the covariant and contravariant basis sets of a space surface can be obtained from each other? State this in words and mathematically defining all the symbols involved.

3.48 What is the significance of the following relation involving the covariant and contravariant surface basis vectors and the Kronecker delta: $\mathbf{E}_\alpha \cdot \mathbf{E}^\beta = \delta_\alpha^\beta$?

3.49 Define, mathematically, the coefficients of the surface metric tensor in terms of the surface covariant basis vectors in Euclidean and Riemannian spaces.

3.50 Give the fundamental relation that provides the important link between the metric tensor of a surface and the metric tensor of its enveloping space.

3.51 Find the surface basis vectors, $\mathbf{E}_1$ and $\mathbf{E}_2$, and the coefficients of the first fundamental form of a surface parameterized by: $\mathbf{r}(u,v) = (au\cos v, bu\sin v, cu^2)$ where $a, b, c$ are constants.

3.52 Find, symbolically, the first fundamental form of a cylinder represented by: $\mathbf{r}(u,v) = (f_1(u), f_2(u), v)$ where $f_1$ and $f_2$ are continuous functions of the given coordinate.

3.53 Given the fact that for 3D Cartesian systems: $I_S = (dx^1)^2 + (dx^2)^2 + (dx^3)^2$ plus the transformation equations from spherical to Cartesian coordinates in 3D, derive $I_S$ for spherical coordinate systems.

3.54 Given the fact that for 3D Cartesian systems: $I_S = (dx^1)^2 + (dx^2)^2 + (dx^3)^2$ plus the transformation equations between general curvilinear and Cartesian coordinate systems in 3D, prove that for general curvilinear systems: $I_S = a_{\alpha\beta}du^\alpha du^\beta$.

3.55 Define, mathematically, each of the following vectors: $x_\alpha^1$, $x_\alpha^2$, $x_\alpha^3$, $x_1^i$, $x_2^i$. Also, discuss their attributes as space and surface vectors and their variance type.

3.56 Discuss how a surface vector can also be considered as a space vector stating the mathematical link between the surface and space representations. Are these representations equivalent? If so, how?

3.57 Write down, using tensor notation, the mathematical relation that correlates the surface basis vectors to the unit normal vector to the surface.

3.58 What is the significance of the following relation which involves the surface contravariant and covariant metric tensors and the Kronecker delta: $a^{\alpha\gamma}a_{\gamma\beta} = \delta_\beta^\alpha$?

3.59 Give the matrix $[a^{\alpha\beta}]$ that represents the contravariant form of the surface metric tensor in terms of the coefficients of the first fundamental form.

3.60 Give the matrix $[a_\beta^\alpha]$ that represents the mixed form of the surface metric tensor.

3.61 Write down the relations that represent the transformation between unbarred and barred surface coordinate systems.

3.62 How the determinants of the surface metric tensor of two transformed coordinate systems of a given surface are linked?

3.63 Express the Christoffel symbols of the first kind $[\alpha\beta, \gamma]$ for a surface in terms of the surface covariant basis vectors and their partial derivatives.

3.64 Derive the mathematical relation between the partial derivative of the surface metric tensor $\partial_\alpha a_{\beta\gamma}$ and the Christoffel symbols of the first kind for the surface.

3.65 How the coefficients of the surface metric tensor will be affected by scaling the surface up or down by a constant positive scalar factor?

3.66 Give the covariant and contravariant types of the surface metric tensor for a Monge patch of the form $\mathbf{r}(u,v) = (u, v, f(u,v))$.

3.67 Discuss how the concept of "length of straight segment" is extended to the length of a polygonal arc. Also discuss how the concept of "length of polygonal arc" is extended to the length of an arc of a twisted space curve.

3.68 Derive the following relation which links the length of an element of arc of a curve residing on a 2D surface to the covariant metric tensor of the surface: $(ds)^2 = a_{\alpha\beta} du^\alpha du^\beta$.

3.69 Is the length of a surface curve an intrinsic or extrinsic property and why?

3.70 Give the formula for the length, $L$, of a segment of a $t$-parameterized surface curve in terms of the coefficients of the first fundamental form of the surface.

3.71 Using the metric tensor, verify the following relation where $f$ represents a Monge patch of the form $\mathbf{r}(u,v) = (u, v, f(u,v))$:

$$ds = \sqrt{(1+f_u^2)\,dudu + 2f_u f_v\, dudv + (1+f_v^2)\,dvdv}$$

3.72 Develop an analytical expression for the length of an element of arc, $ds$, of the catenary parameterized by Eqs. 28-29.

3.73 Discuss how the concept of "area of polygonal plane fragment" is extended to the area of a surface consisting of polygonal plane fragments. Also discuss how the concept of "area of surface made of polygonal plane fragments" is extended to the area of a generalized twisted space surface.

3.74 Derive the mathematical expression for the area of an infinitesimal element of a surface and the expression for the area of a surface patch.

3.75 Derive Eq. 208 for the surface area of a Monge patch of the form $\mathbf{r}(u,v) = (u, v, f(u,v))$.

3.76 A cone is represented parametrically in a 3D space by: $\mathbf{r}(\rho,\phi) = (\rho\cos\phi, \rho\sin\phi, c\rho)$ where $\rho, \phi$ are polar coordinates ($\rho \geq 0$ and $0 \leq \phi < 2\pi$) and $c$ is a positive constant. Find the area of the part of the cone corresponding to $0 \leq \rho \leq A$ where $A$ is a given positive constant.

3.77 Derive the mathematical expression: $\cos\theta = g_{ij}A^i B^j$ for the angle $\theta$ between two unit surface vectors, $\mathbf{A}$ and $\mathbf{B}$, where $g_{ij}$ is the space covariant metric tensor. Also give the mathematical expression of $\sin\theta$ for the angle between $\mathbf{A}$ and $\mathbf{B}$.

3.78 Give two mathematical expressions for the coefficients of the surface covariant curvature tensor $b_{\alpha\beta}$.

3.79 Using a spherical coordinates representation, find the determinant of the surface curvature tensor of a sphere.

3.80 Show that $\partial_\beta \mathbf{E}_\alpha = \partial_\alpha \mathbf{E}_\beta$.

3.81 Prove that if two surfaces, $S_1$ and $S_2$, are mapped isometrically one on the other then they have identical first fundamental form coefficients at their corresponding points, i.e. $E_1 = E_2$, $F_1 = F_2$ and $G_1 = G_2$.

3.82 Give the matrix $[b^{\alpha\beta}]$ which represents the contravariant form of the surface curvature tensor in terms of the coefficients of the first and second fundamental forms.

3.83 Discuss and justify the relation $\bar{b} = J^2 b$ explaining all the symbols involved.

3.84 What are the other symbols used by some authors to label the coefficients of the second fundamental form $e, f, g$? Discuss the advantages and disadvantages of using each one of these sets of symbols. Also, write the mathematical expression for the

## 3.12 Exercises

second fundamental form using the alternative symbols.

3.85 How can we obtain the mixed form of the surface curvature tensor $b^\alpha_\beta$ from the covariant form of this tensor $b_{\alpha\beta}$?

3.86 Express the mean curvature $H$ and the Gaussian curvature $K$ of a surface in terms of the mixed form of the surface curvature tensor $b^\alpha_\beta$.

3.87 Explain, in details, all the symbols and notations involved in the following relation:

$$b^\alpha_\gamma b_{\beta\delta} - b^\alpha_\delta b_{\beta\gamma} = \frac{\partial \Gamma^\alpha_{\beta\delta}}{\partial \gamma} - \frac{\partial \Gamma^\alpha_{\beta\gamma}}{\partial \delta} + \Gamma^\omega_{\beta\delta}\Gamma^\alpha_{\omega\gamma} - \Gamma^\omega_{\beta\gamma}\Gamma^\alpha_{\omega\delta}$$

3.88 Write down the matrix form of the surface covariant curvature tensor for a Monge patch of the form $\mathbf{r}(u,v) = (u, v, f(u,v))$.

3.89 What is the relation between the Riemann-Christoffel curvature tensor of the second kind and the curvature tensor of a surface?

3.90 Give all the coefficients of the Riemann-Christoffel curvature tensor and the curvature tensor for a plane surface.

3.91 On a space surface, how many independent non-vanishing components the Riemann-Christoffel curvature tensor possesses?

3.92 Explain the relation between the coefficients of the metric tensor of a surface and its first fundamental form stating the necessary equations.

3.93 Derive the mathematical formula for $I_S$ in terms of the coefficients of the first fundamental form.

3.94 Express the determinant of the surface covariant metric tensor as a function of the coefficients of the first fundamental form.

3.95 Express $E, F, G$ as dot products of the covariant basis vectors of the surface and relate this to the space metric tensor.

3.96 Does the first fundamental form provide a unique characterization of the space surface as seen internally by a 2D inhabitant? Explain why.

3.97 Does the first fundamental form provide a unique characterization of the space surface as seen from the external ambient space? Explain why.

3.98 State the mathematical conditions for the provision of positive definiteness of the first fundamental form.

3.99 Give the mathematical conditions that apply to the coefficients of the covariant metric tensor of two isometric surfaces at their corresponding points.

3.100 Explain how the second fundamental form characterizes the surface from the ambient space perspective and how the unit normal vector to the surface is employed in this characterization.

3.101 Express the determinant of the surface covariant curvature tensor as a function of the coefficients of the second fundamental form.

3.102 Derive the mathematical relation of the second fundamental form $II_S$ in terms of the coefficients $e, f, g$. Also provide the main mathematical definitions for the coefficients $e, f, g$ in terms of the surface basis vectors and the unit normal vector to the surface.

3.103 What is the relation between the second fundamental form $II_S$ and the second order differential of the position vector $d^2\mathbf{r}$ of a surface?

## 3.12 Exercises

3.104 State the mathematical relations between the coefficients of the second fundamental form and the coefficients of the surface covariant curvature tensor.

3.105 Express, in full tensor notation, the second fundamental form in terms of the coefficients of the surface covariant curvature tensor and link this to the expression involving the coefficients $e, f, g$.

3.106 Explain how the following relation provides a bridge between the first and second fundamental forms: $II_S = \kappa_n I_S$.

3.107 Derive the following relation where $f$ is a functional representation of a Monge patch of the form $\mathbf{r}(u,v) = (u, v, f(u,v))$:

$$II_S = \frac{f_{uu} du du + 2 f_{uv} du dv + f_{vv} dv dv}{\sqrt{1 + f_u^2 + f_v^2}}$$

3.108 Discuss how the first and second fundamental forms represent the intrinsic and extrinsic geometry of the surface.

3.109 If two surfaces have identical first and second fundamental forms, should they be congruent?

3.110 What are the compatibility conditions linking the first and second fundamental forms which are needed to fully identify a surface associated with specific first and second fundamental forms and secure its existence?

3.111 State, using mathematical technical terms, the fundamental theorem of space surfaces.

3.112 Give a brief definition of Dupin indicatrix and state its significance and usage in differential geometry.

3.113 What is the shape of Dupin indicatrix at elliptic, parabolic and hyperbolic points on a smooth surface? What is the shape of Dupin indicatrix at flat points?

3.114 Make a simple sketch to illustrate Dupin indicatrix at an elliptic point, a parabolic point and a hyperbolic point on a surface marking the two principal directions in each case.

3.115 Write down the mathematical expression for the third fundamental form $III_S$ in terms of the unit normal vector to the surface and in terms of the coefficients $c_{\alpha\beta}$.

3.116 Express the coefficients of the third fundamental form as a function of the coefficients of the surface metric and curvature tensors.

3.117 Explain all the symbols involved in the following equation: $KI_S - 2H\,II_S + III_S = 0$.

3.118 Derive the equation in the last question using the Weingarten equations.

3.119 Starting from the equation: $Ka_{\alpha\beta} - 2Hb_{\alpha\beta} + c_{\alpha\beta} = 0$, derive, with full explanation, the following relation: $\operatorname{tr}\left(c_\alpha^\beta\right) = 4H^2 - 2K$.

3.120 Explain the correspondence between the Frenet-Serret formulae for space curves and the equations of Gauss and Weingarten for space surfaces.

3.121 Why the partial derivatives of the surface basis vectors, $\mathbf{E}_1$ and $\mathbf{E}_2$, and the unit normal vector to the surface, $\mathbf{n}$, with respect to the surface coordinates, $u^1$ and $u^2$, can be expressed as combinations of these vectors?

3.122 Prove Eq. 282 using the Weingarten equations.

3.123 Write the derivatives of the surface basis vectors (i.e. $\partial_\beta \mathbf{E}_\alpha$) in terms of the surface vectors (i.e. $\mathbf{E}_\gamma$ and $\mathbf{n}$) in their vector and tensor forms.

## 3.12 Exercises

3.124 What is the essence of the equations of Weingarten? Provide qualitative and quantitative descriptions of these equations. Also write these equations in a matrix form involving the covariant curvature tensor and the contravariant metric tensor of the surface.

3.125 Derive Weingarten equations for a Monge patch of the form $\mathbf{r}(u,v) = (u, v, f(u,v))$.

3.126 Express the partial derivatives of $\mathbf{n}$ with respect to the surface coordinates in terms of the Gaussian and mean curvatures and the coefficients of the first and second fundamental forms.

3.127 Give the equations of Gauss for a Monge patch of the form $\mathbf{r}(u,v) = (u, v, f(u,v))$.

3.128 State, using full tensor notation, the equations of Codazzi-Mainardi explaining all the symbols involved.

3.129 Derive, using the Codazzi-Mainardi equations, the following relation: $b_{\alpha\beta;\gamma} = b_{\alpha\gamma;\beta}$.

3.130 Explain how the relation in the previous question indicates that there are only two independent components for the Codazzi-Mainardi equations. What are these two independent components?

3.131 Describe sphere mapping in qualitative and technical terms. Also, explain the meaning of the following equation:

$$\lim_{\mathfrak{R} \to P} \frac{\sigma(\bar{\mathfrak{R}})}{\sigma(\mathfrak{R})} = |K_P|$$

3.132 Prove the theorem represented by the equation in the previous question.

3.133 Prove that at a given point $P$ on a surface with $K \neq 0$, there exists a neighborhood $\mathcal{N}$ of $P$ where an injective mapping can be established between $\mathcal{N}$ and its spherical image $\bar{\mathcal{N}}$.

3.134 State one of the global theorems of space surface and explain why it is global.

# Chapter 4
# Curvature

"Curvature" is a property of both curves and surfaces at a given point which is determined by the shape of the curve or surface at that point. There are also global characteristics of curvature like total curvature $K_t$ (see § 4.8) of a surface but they are based in general on the local characterization of curvature at individual points. The curvature also has intrinsic as well as extrinsic attributes and hence it characterizes the manifold internally as seen by an inhabitant of the manifold and externally as seen by an outsider. In this chapter, we investigate this property in its general meaning and examine the main parameters used to describe curvature and quantify it focusing on space surfaces and curves embedded in such surfaces. The materials are largely based on a 3D flat ambient space coordinated by a Cartesian orthonormal system.

## 4.1 Curvature Vector

At a given point $P$ on a surface $S$, a plane containing the vector $\mathbf{n}$, which is the normal unit vector to the surface at $P$, intersects the surface in a surface curve $C$ having a tangent vector $\mathbf{t}$ at $P$. The curve $C$ is called the normal section of $S$ at $P$ in the direction of $\mathbf{t}$. The principal normal vector $\mathbf{N}$ of a normal section at $P$ is collinear with the unit normal vector $\mathbf{n}$ at $P$. We note that for a normal section, $\mathbf{N}$ and $\mathbf{n}$ can be parallel or anti-parallel since on an orientable surface the vector $\mathbf{n}$ can have one of two possible directions. On the other hand, a surface curve passing through $P$ in the direction of $\mathbf{t}$ may not be a normal section and hence the vectors $\mathbf{N}$ and $\mathbf{n}$ at $P$ have different orientations. In this context we remark that we are considering here the part of the curve in the neighborhood of the point $P$ as part of a normal section or not, and not necessarily the whole curve.

As explained before, a space curve $C$ can be parameterized by $s$, representing the distance traversed along $C$, and hence the curve is defined by the position vector $\mathbf{r}(s)$. At a given point $P$ on $C$, the vector $\mathbf{T} = \frac{d\mathbf{r}}{ds}$ is a unit vector tangent to $C$ at $P$ in the direction of increasing $s$. When $C$ is embedded in a surface, $\mathbf{T}$ will be contained in the tangent plane to the surface at $P$, as explained previously in § 2.2 and 3.1. We note that the vector $\mathbf{T}$ can be parallel or anti-parallel to the aforementioned vector $\mathbf{t}$. We chose to introduce $\mathbf{t}$ and define it in this way to be more general since the curve orientation can be in one direction or the other and hence $\mathbf{T}$ and $\mathbf{t}$ can be parallel or anti-parallel. Moreover, $\mathbf{t}$ is not necessarily of unit length or based on a natural parameterization of the curve.

The curvature vector of $C$ at $P$, which is orthogonal to $\mathbf{T}$, is defined by:

$$\mathbf{K} = \frac{d\mathbf{T}}{ds} \tag{304}$$

where $\mathbf{K}$, which is the uppercase Greek letter kappa, symbolizes the curvature vector.

## 4.1 Curvature Vector

The curvature $\kappa$ of $C$ at $P$ (which is defined previously in § 2.2) is the magnitude of the curvature vector, that is: $\kappa = |\mathbf{K}|$, and the radius of curvature when $\kappa \neq 0$ is its reciprocal, i.e. $R_\kappa = \frac{1}{\kappa}$, which is the radius of the osculating circle of $C$ at $P$ (see § 2.6). The curvature vector can therefore be expressed as:

$$\mathbf{K} = |\mathbf{K}| \frac{\mathbf{K}}{|\mathbf{K}|} = \kappa \mathbf{N} \qquad (305)$$

where $\mathbf{N}$ is the principal normal vector of the curve $C$ at $P$ as defined previously in § 2.2.

The curvature vector of a surface curve is independent of the orientation and parameterization of the surface and the curve. However, the curvature vector at a particular point is determined by the local shape of the curve, which is partly determined by the position on the surface and the tangential direction to the surface at that position, and hence the curvature vector depends on the surface point and tangential direction, as well as on other factors.

A point on the curve at which the curvature vector $\mathbf{K}$ vanishes, and hence $\kappa = 0$, is called inflection point. At such a point, the radius of curvature is infinite and the principal normal vector $\mathbf{N}$ and the osculating circle are not defined. However, since it is usually assumed that the curve is of class $C^2$, the curvature vector varies smoothly and hence at isolated points of inflection on such a curve, $\mathbf{N}$ may be defined in such a way to ensure continuity when this is possible, which is not always the case.

On introducing a new unit vector which is orthogonal to both $\mathbf{n}$ and $\mathbf{T}$ and defined by the following cross product:

$$\mathbf{u} = \mathbf{n} \times \mathbf{T} \qquad (306)$$

the curvature vector, which lies in the plane spanned by $\mathbf{n}$ and $\mathbf{u}$, can then be resolved in the $\mathbf{n}$ and $\mathbf{u}$ directions, which represent the normal and tangential directions to the surface at the given point, as:

$$\mathbf{K} = \mathbf{K}_n + \mathbf{K}_g = \kappa_n \mathbf{n} + \kappa_g \mathbf{u} \qquad (307)$$

where $\mathbf{K}_n$ and $\mathbf{K}_g$ are the normal and geodesic components of the curvature vector $\mathbf{K}$, while $\kappa_n$ and $\kappa_g$ are the normal and geodesic curvatures of the curve at the given point respectively.

On dot producting both sides of Eq. 307 with $\mathbf{n}$ and $\mathbf{u}$ in turn, noting that $\mathbf{n}$ and $\mathbf{u}$ are orthogonal unit vectors, the normal and geodesic curvatures can be obtained, that is:

$$\kappa_n = \mathbf{n} \cdot \mathbf{K} = -\mathbf{T} \cdot \frac{d\mathbf{n}}{ds} = -\frac{d\mathbf{r}}{ds} \cdot \frac{d\mathbf{n}}{ds} \qquad (308)$$

$$\kappa_g = \mathbf{u} \cdot \mathbf{K} = \mathbf{u} \cdot \frac{d\mathbf{T}}{ds} = (\mathbf{n} \times \mathbf{T}) \cdot \frac{d\mathbf{T}}{ds} \qquad (309)$$

The second equality of Eq. 308 is based on the fact that $\mathbf{T}$ and $\mathbf{n}$ are orthogonal (since $\mathbf{T}$ is tangent to the surface while $\mathbf{n}$ is normal to the surface); therefore by the product rule of differentiation we have:

$$\frac{d(\mathbf{n} \cdot \mathbf{T})}{ds} = \frac{d(0)}{ds} = 0 = \mathbf{n} \cdot \frac{d\mathbf{T}}{ds} + \frac{d\mathbf{n}}{ds} \cdot \mathbf{T} = \mathbf{n} \cdot \mathbf{K} + \mathbf{T} \cdot \frac{d\mathbf{n}}{ds} \qquad (310)$$

## 4.1 Curvature Vector

and hence: $\mathbf{n} \cdot \mathbf{K} = -\mathbf{T} \cdot \frac{d\mathbf{n}}{ds}$. The other equalities in Eqs. 308 and 309 are based on definitions which have been given previously.

The vector $\mathbf{u}$, which is a unit vector normal to the curve $C$, is called the geodesic normal vector. This vector is the normalized projection of $\mathbf{K}$ onto the tangent space of the surface and hence it is contained in the tangent plane of the surface at the given point. Also, because the vector $\mathbf{u}$ is orthogonal to the curve $C$ at $P$ (since $\mathbf{u}$ is orthogonal to $\mathbf{T}$ by the cross product), it is contained in the normal plane of $C$ at $P$. Hence, $\mathbf{u}$ occurs at the intersection of the tangent plane of the surface and the normal plane of the curve which correspond to the given point.

While the normal curvature $\kappa_n$ is an extrinsic property, since it depends on the first and second fundamental form coefficients, as seen in § 3.6 (Eq. 260) and as will be seen in 4.2, the geodesic curvature $\kappa_g$ in an intrinsic property as it depends only on the first fundamental form coefficients and their derivatives (see § 4.3). We note that the triad $(\mathbf{n}, \mathbf{T}, \mathbf{u})$ is another moving frame which is in common use in differential geometry in addition to the curve-based Frenet frame $(\mathbf{T}, \mathbf{N}, \mathbf{B})$ and the surface-based frame $(\mathbf{E}_1, \mathbf{E}_2, \mathbf{n})$.

Regarding the relation between the principal normal vector of the curve $\mathbf{N}$, and the unit normal vector to the surface $\mathbf{n}$, let $C$ be a curve on a sufficiently smooth surface $S$. If $\phi$ is the angle between the vector $\mathbf{N}$ of $C$ at a given point $P$ and the vector $\mathbf{n}$ of $S$ at $P$ then we have:

$$\cos\phi = \mathbf{n} \cdot \mathbf{N} \qquad (311)$$

Accordingly, the normal and geodesic curvatures, $\kappa_n$ and $\kappa_g$, of $C$ at $P$ are given by:

$$\kappa_n = \kappa \cos\phi \qquad (312)$$
$$\kappa_g = \kappa \sin\phi \qquad (313)$$

where $\kappa$ is the curvature of $C$ at $P$ as defined previously (see § 2.2 and the previous parts of the present section). In fact, Eq. 312 can be easily obtained by combining Eq. 308 with Eq. 305, while Eq. 313 can be similarly obtained by combining Eq. 309 with Eq. 305, that is:

$$\kappa_n = \mathbf{n} \cdot \mathbf{K} = \mathbf{n} \cdot (\kappa \mathbf{N}) = \kappa (\mathbf{n} \cdot \mathbf{N}) = \kappa \cos\phi \qquad (314)$$
$$\kappa_g = \mathbf{u} \cdot \mathbf{K} = \mathbf{u} \cdot (\kappa \mathbf{N}) = \kappa (\mathbf{u} \cdot \mathbf{N}) = \kappa \cos\left(\frac{\pi}{2} - \phi\right) = \kappa \sin\phi \qquad (315)$$

According to the theorem of Meusnier, if $P$ is a given point on a sufficiently smooth surface $S$, then all curves on $S$ that pass through $P$ with the same tangent direction at $P$ have the same normal curvature at $P$. The theorem of Meusnier may also be stated in this context as: the curvature of any surface curve at a given point $P$ on the curve is equal in magnitude to the curvature of the normal section which is tangent to the curve at $P$ divided by the cosine of the angle between the principal normal vector to the curve at $P$ and the normal vector to the surface at $P$. This version of the theorem is based on Eq. 312 plus the fact that the normal curvature $\kappa_n$ of a normal section is equal in magnitude to its curvature $\kappa$. Also, we should exclude from this version the case of having $\cos\phi = 0$ when the curvature vector of the curve is tangent to the surface. More details about Meusnier theorem will be given in § 4.2.1.

## 4.2 Normal Curvature

Using the first and second fundamental forms, given by Eqs. 233 and 244, the normal curvature at a given point on the surface in the $\frac{du^2}{du^1}$ direction can be expressed as the following quotient of the second fundamental form involving the coefficients of the surface covariant curvature tensor to the first fundamental form involving the coefficients of the surface covariant metric tensor (see Eq. 259):

$$\kappa_n = \frac{II_S}{I_S} = \frac{e(du^1)^2 + 2f\, du^1 du^2 + g(du^2)^2}{E(du^1)^2 + 2F\, du^1 du^2 + G(du^2)^2} = \frac{b_{\alpha\beta} du^\alpha du^\beta}{a_{\gamma\delta} du^\gamma du^\delta} \tag{316}$$

where the symbols are as explained before, and the last part is to be interpreted as the sum of the terms in the numerator divided by the sum of the terms in the denominator. This equation can be obtained as follows:

$$\kappa_n = -\mathbf{T} \cdot \frac{d\mathbf{n}}{ds} \qquad \text{(Eq. 308)} \tag{317}$$

$$= -\left(\frac{\partial \mathbf{r}}{\partial u^\alpha} \frac{du^\alpha}{ds}\right) \cdot \left(\frac{\partial \mathbf{n}}{\partial u^\beta} \frac{du^\beta}{ds}\right) \qquad \text{(Eq. 106)}$$

$$= -\left(\frac{\partial \mathbf{r}}{\partial u^\alpha} \cdot \frac{\partial \mathbf{n}}{\partial u^\beta}\right) \frac{du^\alpha}{ds} \frac{du^\beta}{ds}$$

$$= \frac{-\left(\frac{\partial \mathbf{r}}{\partial u^\alpha} \cdot \frac{\partial \mathbf{n}}{\partial u^\beta}\right) du^\alpha du^\beta}{ds\, ds}$$

$$= \frac{b_{\alpha\beta} du^\alpha du^\beta}{(ds)^2} \qquad \text{(Eq. 213)}$$

$$= \frac{b_{\alpha\beta} du^\alpha du^\beta}{a_{\gamma\delta} du^\gamma du^\delta} \qquad \text{(Eq. 233)}$$

From the previous statements plus Eq. 307, it can be seen that the normal component of the curvature vector may also be given by:

$$\mathbf{K}_n = \left[e\left(\frac{du^1}{ds}\right)^2 + 2f\frac{du^1}{ds}\frac{du^2}{ds} + g\left(\frac{du^2}{ds}\right)^2\right] \mathbf{n} \tag{318}$$

Also, from Eq. 316 it can be seen that the sign of the normal curvature $\kappa_n$ (i.e. being greater than, less than or equal to zero) is determined solely by the sign of the second fundamental form since the first fundamental form is positive definite. As seen before (see § 3.6), the sign of the second fundamental form depends on the surface orientation which is determined by the direction of $\mathbf{n}$. Therefore, the sign of $\kappa_n$ depends on the surface orientation. Apart from this, the sign of the second fundamental form is related to the sign of the determinant of the surface covariant curvature tensor $b$, and hence the sign of $\kappa_n$ is also related to the sign of $b$.

As given earlier, all surface curves passing through a given point $P$ on a surface and have the same tangent line at $P$ have identical normal curvature at $P$. Hence, the normal

## 4.2 Normal Curvature

curvature is a property of the surface at a given point and in a given direction and not only a property of the curve. The normal curvature $\kappa_n$ of a given normal section $C$ of a surface at a particular point $P$ is equal in magnitude to the curvature $\kappa$ of $C$ at $P$, i.e. $|\kappa_n| = \kappa$. This can be explained by the fact that the normal vector $\mathbf{n}$ to the surface at $P$ is collinear with the principal normal vector $\mathbf{N}$ of $C$ at $P$ so there is only a normal component to the curvature vector with no tangential geodesic component.

The normal curvature of a surface at a given point and in a given spatially-fixed tangential direction is an invariant property with respect to change of parameterization and representation of the surface apart from its sign which is dependent on the choice of the direction of the unit normal vector $\mathbf{n}$ to the surface, as seen earlier. The normal curvature is an extrinsic property, since it necessarily depends on the coefficients of the second fundamental form, and hence it cannot be expressed purely in terms of the coefficients of the first fundamental form.

At flat points on a surface, $\kappa_n = 0$ in all directions. At elliptic points, $\kappa_n \neq 0$ in any direction and it has the same sign in all directions. At parabolic points, $\kappa_n$ has the same sign in all directions except the direction in which the second fundamental form vanishes where $\kappa_n = 0$. At hyperbolic points, $\kappa_n$ is negative, positive and zero depending on the direction. For the definition of flat, elliptic, parabolic and hyperbolic points and their significance, the reader is referred to § 4.9.

In any two orthogonal tangential directions at a given point $P$ on a sufficiently smooth surface, the sum of the normal curvatures corresponding to these directions at $P$ is constant. At any point $P$ of a sufficiently smooth surface $S$, there exists a paraboloid (elliptic or hyperbolic) which is tangent at its vertex to the tangent plane of $S$ at $P$ such that the normal curvature of the paraboloid in a given direction at $P$ is equal to the normal curvature of $S$ at $P$ in that direction. This paraboloid takes the degenerate form of a plane at flat points and a parabolic cylinder at parabolic points (see § 4.9).

The normal curvatures of the surface curves at a given point $P$ in the directions of the $u$ and $v$ coordinate curves are given respectively by:

$$\kappa_{nu} = \frac{b_{11}}{a_{11}} = \frac{e}{E} \qquad (319)$$

$$\kappa_{nv} = \frac{b_{22}}{a_{22}} = \frac{g}{G} \qquad (320)$$

where $\kappa_{nu}$ and $\kappa_{nv}$ are the normal curvatures in the directions of the $u$ and $v$ coordinate curves, the indexed $a$ and $b$ are the coefficients of the covariant metric tensor and the covariant curvature tensor and where these are evaluated at $P$. This can be seen, for example, from Eq. 316 where the last two terms in the sums will vanish for the $u^1$ coordinate curve since $du^2 = 0$ while the first two terms in the sums will vanish for the $u^2$ coordinate curve since $du^1 = 0$. We note that $du^1 \neq 0$ on the $u^1$ coordinate curve and $du^2 \neq 0$ on the $u^2$ coordinate curve. As we will see in § 4.4, at each non-umbilical point $P$ (refer to § 4.10) of a sufficiently smooth surface there are two perpendicular directions along which the normal curvature of the surface at $P$ takes its maximum and minimum values of all the normal curvature values at $P$.

The necessary and sufficient condition for a given point $P$ on a sufficiently smooth surface $S$ to be umbilical point is that the coefficients of the first and second fundamental forms of the surface at $P$ are proportional, that is:

$$\frac{e}{E} = \frac{f}{F} = \frac{g}{G} (= \kappa_n) \qquad (321)$$

where $E, F, G, e, f, g$ are the coefficients of the first and second fundamental forms at $P$, and $\kappa_n$ is the normal curvature of $S$ at $P$ in any direction. This can be seen from Eq. 316 where $\kappa_n$ in this case becomes independent of the direction since by taking out the common proportionality factor the expression will be reduced to: $\kappa_n = c$ where $c$ is the proportionality factor. More explicitly, according to Eq. 321: $e = cE$, $f = cF$ and $g = cG$ where $c$ is the common proportionality factor, and hence from Eq. 316 we obtain: $\kappa_n = c \times 1 = c$.

As seen before, the curvature $\kappa$ and the normal curvature $\kappa_n$ of a surface curve at a given point $P$ on the curve are related by:

$$\kappa_n = \kappa \cos \phi \qquad (322)$$

where $\phi$ is the angle between the principal normal vector $\mathbf{N}$ of the curve at $P$ and the unit normal vector $\mathbf{n}$ of the surface at $P$.

At every point on a sphere and in any direction, the normal curvature is constant given by: $|\kappa_n| = \frac{1}{R}$ where $R$ is the sphere radius. At any point $P$ on a sphere, any surface curve $C$ passing through $P$ in any direction is a normal section iff $C$ is a great circle. All these great circles have constant curvature $\kappa$ and normal curvature $\kappa_n$ which are both equal in magnitude to $\frac{1}{R}$ where $R$ is the sphere radius. We remark that the great circles of a sphere are the plane sections formed by the intersection of the sphere with the planes passing through the center of the sphere.

## 4.2.1 Meusnier Theorem

According to the theorem of Meusnier, all surface curves passing through a given point $P$ on a surface and have the same tangential non-asymptotic direction (see § 5.9) at $P$ have identical normal curvature which is the normal curvature $\kappa_n$ (and the curvature $\kappa$ considering the magnitude) of the normal section at $P$ in the given direction. Moreover, the osculating circles of these curves lie on a sphere $S_s$ with radius $\frac{1}{\kappa}$ and with center at $\mathbf{r}_C = \mathbf{r}_P + \frac{\mathbf{N}}{\kappa}$ where $\kappa$ (which is equal in magnitude to $\kappa_n$) is the curvature of the normal section at $P$, $\mathbf{N}$ is the principal normal vector of the normal section at $P$ and $\mathbf{r}_P$ is the position vector of $P$. As a consequence of this theorem, we have:
1. The center of the sphere $S_s$ is the center of curvature (see § 2.6) of the normal section at $P$ in the given direction.
2. These curves are characterized by being tangent to the normal section at $P$ in the given direction and by being plane sections of the surface with shared tangent direction at $P$.[16]

---

[16] We are considering here the part of these curves in the neighborhood of the point $P$ on the surface.

3. The osculating circles of these curves are the intersection of the sphere $S_s$ with the osculating planes of these curves at $P$.
4. The sphere $S_s$ is tangent to the tangent plane of the surface at $P$.

The theorem of Meusnier may also be stated as follows: the center of curvature of a surface curve at a given point $P$ on the curve is obtained by orthogonal projection of the center of curvature of the normal section, which is tangent to the curve at $P$, on the osculating plane of the curve.

## 4.3 Geodesic Curvature

As described earlier (see § 4.1), the curvature vector $\mathbf{K}$ of a surface curve lies in a plane perpendicular to the tangent vector $\mathbf{T}$ and it can be resolved into a normal component $\mathbf{K}_n = \kappa_n \mathbf{n}$ and a geodesic component $\mathbf{K}_g = \kappa_g \mathbf{u}$ where the normal and geodesic curvatures, $\kappa_n$ and $\kappa_g$, are given by Eqs. 308 and 309. The geodesic component $\mathbf{K}_g$ of the curvature vector $\mathbf{K}$ of a surface curve at a given point $P$ and in a given direction is the projection of $\mathbf{K}$ onto the tangent space $T_P S$ of the surface at $P$. This geodesic component of the curvature vector of a surface curve is given by:

$$\mathbf{K}_g = \kappa_g \mathbf{u} = \left( \frac{d^2 u^1}{ds^2} + \Gamma^1_{\alpha\beta} \frac{du^\alpha}{ds} \frac{du^\beta}{ds} \right) \mathbf{E}_1 + \left( \frac{d^2 u^2}{ds^2} + \Gamma^2_{\alpha\beta} \frac{du^\alpha}{ds} \frac{du^\beta}{ds} \right) \mathbf{E}_2 \qquad (323)$$

where the Christoffel symbols are derived from the surface metric. Since $\mathbf{u} = \mathbf{n} \times \mathbf{T}$, the sense of the geodesic curvature vector depends on the orientation of the surface and the orientation of the curve. The geodesic component of the curvature vector may also be given by the following expression:

$$\mathbf{K}_g = \left[ \mathbf{n} \times \left( \frac{\partial \mathbf{E}_1}{\partial u^1} \left( \frac{du^1}{ds} \right)^2 + 2 \frac{\partial \mathbf{E}_1}{\partial u^2} \frac{du^1}{ds} \frac{du^2}{ds} + \frac{\partial \mathbf{E}_2}{\partial u^2} \left( \frac{du^2}{ds} \right)^2 \right) \right] \times \mathbf{n} \\ + \mathbf{E}_1 \frac{d^2 u^1}{ds^2} + \mathbf{E}_2 \frac{d^2 u^2}{ds^2} \qquad (324)$$

where the symbols are as explained before.

As well as the previously developed expressions for $\kappa_g$ (see Eqs. 309 and 313), it can be shown that for a naturally parameterized curve the geodesic curvature $\kappa_g$ is also given by:

$$\kappa_g = \sqrt{a} \left[ \Gamma^2_{11} \left( \frac{du^1}{ds} \right)^3 + (2\Gamma^2_{12} - \Gamma^1_{11}) \left( \frac{du^1}{ds} \right)^2 \frac{du^2}{ds} + \\ (\Gamma^2_{22} - 2\Gamma^1_{12}) \frac{du^1}{ds} \left( \frac{du^2}{ds} \right)^2 - \Gamma^1_{22} \left( \frac{du^2}{ds} \right)^3 + \frac{du^1}{ds} \frac{d^2 u^2}{ds^2} - \frac{d^2 u^1}{ds^2} \frac{du^2}{ds} \right] \qquad (325)$$

where the Christoffel symbols are derived from the surface metric and $a = EG - F^2$ is the determinant of the surface covariant metric tensor. While the curvature $\kappa$ and the normal curvature $\kappa_n$ are extrinsic properties of the surface, the geodesic curvature $\kappa_g$ is an intrinsic property. This can be seen, for example, from Eq. 325.

## 4.3 Geodesic Curvature

On the $u^1$ coordinate curves, $\frac{du^2}{ds} = 0$ and $\frac{du^1}{ds} = \frac{1}{\sqrt{E}}$.[17] Hence, Eq. 325 will simplify to:

$$\kappa_{gu} = \sqrt{a}\, \Gamma_{11}^2 \left(\frac{du^1}{ds}\right)^3 = \frac{\sqrt{a}}{E^{3/2}} \Gamma_{11}^2 \tag{326}$$

where $\kappa_{gu}$ is the geodesic curvature of the $u^1$ coordinate curve. The last formula will be simplified further if the $u^1$ and $u^2$ coordinate curves are orthogonal, since in this case $F = 0$ and $\Gamma_{11}^2 = -\frac{E_v}{2G}$ (see Eq. 74), and the formula will become:

$$\kappa_{gu} = -\frac{E_v}{2E\sqrt{G}} \tag{327}$$

Similarly, on the $u^2$ coordinate curves, $\frac{du^1}{ds} = 0$ and $\frac{du^2}{ds} = \frac{1}{\sqrt{G}}$.[18] Hence, Eq. 325 will simplify to:

$$\kappa_{gv} = -\sqrt{a}\, \Gamma_{22}^1 \left(\frac{du^2}{ds}\right)^3 = -\frac{\sqrt{a}}{G^{3/2}} \Gamma_{22}^1 \tag{328}$$

where $\kappa_{gv}$ is the geodesic curvature of the $u^2$ coordinate curve. The last formula will be simplified further if the $u^1$ and $u^2$ coordinate curves are orthogonal, since in this case $F = 0$ and $\Gamma_{22}^1 = -\frac{G_u}{2E}$ (see Eq. 77), and the formula then becomes:

$$\kappa_{gv} = \frac{G_u}{2G\sqrt{E}} \tag{329}$$

Although the geodesic curvature is an intrinsic property, as can be seen from the above equations, it can also be calculated extrinsically by:

$$\kappa_g = \frac{\ddot{\mathbf{r}} \cdot (\mathbf{n} \times \dot{\mathbf{r}})}{\dot{\mathbf{r}} \cdot \dot{\mathbf{r}}} \tag{330}$$

where the overdots stand for differentiation with respect to a general parameter $t$ of the curve. As discussed previously, some intrinsic properties can also be defined in terms of extrinsic parameters.

As seen before, the curvature $\kappa$ and the geodesic curvature $\kappa_g$ of a surface curve at a given point $P$ on the curve are related by:

$$\kappa_g = \kappa \sin \phi \tag{331}$$

where $\phi$ is the angle between the principal normal vector $\mathbf{N}$ of the curve at $P$ and the unit normal vector $\mathbf{n}$ of the surface at $P$. It should be remarked that the geodesic curvature

---

[17] This may be demonstrated non-rigorously as:

$$\frac{ds}{du^1}\frac{ds}{du^1} = \frac{Edu^1 du^1}{du^1 du^1} = E$$

since on the $u^1$ coordinate curves we have: $I_S = (ds)^2 = E(du^1)^2$ (see Eq. 233). Hence, $\frac{du^1}{ds} = \frac{1}{\sqrt{E}}$.

[18] This can be demonstrated non-rigorously as in the previous footnote by replacing $du^1$ with $du^2$ and $E$ with $G$.

can take any real value: positive, negative or zero. However, there are some details related to the definition of the angle $\phi$ in the last equation and the geodesic normal vector **u** that should be considered.

On a surface patch of class $C^2$ with orthogonal coordinate curves, an $s$-parameterized curve $C$ of class $C^2$ has a geodesic curvature given by:

$$\kappa_g = \frac{d\theta}{ds} + \kappa_{gu}\cos\theta + \kappa_{gv}\sin\theta \qquad (332)$$

where $\kappa_{gu}$ and $\kappa_{gv}$ are the geodesic curvatures of the $u$ and $v$ coordinate curves and $\theta$ is the angle such that:

$$\mathbf{T} = \frac{\mathbf{E}_1}{|\mathbf{E}_1|}\cos\theta + \frac{\mathbf{E}_2}{|\mathbf{E}_2|}\sin\theta \qquad (333)$$

where **T** is the tangent unit vector of $C$, and $\mathbf{E}_1$ and $\mathbf{E}_2$ are the surface basis vectors, and where all the given quantities are evaluated at a given point on the curve. We note that Eq. 332 is known as Liouville formula.

## 4.4 Principal Curvatures and Directions

On rotating the plane containing **n** (i.e. the unit normal vector to the surface at a given point $P$ on the surface) around **n**, the normal section and hence its curvature $\kappa$ and normal curvature $\kappa_n$ at $P$ will vary in general.[19] Based on the previous findings (see § 4.2), the normal curvature $\kappa_n$ (which, for a normal section, is equal in magnitude to its curvature $\kappa$) of the surface at $P$ in a given direction $\lambda$ can be given by:

$$\kappa_n = \frac{e + 2f\lambda + g\lambda^2}{E + 2F\lambda + G\lambda^2} \qquad (334)$$

where $E, F, G, e, f, g$ are the coefficients of the first and second fundamental forms and $\lambda = \frac{du^2}{du^1}$. In fact, Eq. 334 is a variant of Eq. 316 obtained by dividing the numerator and denominator of Eq. 316 by $(du^1)^2$. The directions represented by $\frac{du^2}{du^1}$ are the directions of the tangents to the normal sections at $P$. We note that for ease of notation and expression, symbols like $\frac{du^2}{du^1}$ and $du : dv$ are laxly labeled as directions since a pair like $(du^1, du^2)$ or $(du, dv)$ represents a direction.

The two principal curvatures of the surface at $P$, $\kappa_1$ and $\kappa_2$, which represent respectively the maximum and minimum values of the normal curvature $\kappa_n$ of the surface at $P$ as given by Eq. 334, correspond to the two $\lambda$ roots of the following quadratic equation:

$$(gF - fG)\lambda^2 + (gE - eG)\lambda + (fE - eF) = 0 \qquad (335)$$

where $(gF - fG) \neq 0$. The last equation is obtained by equating the derivative of $\kappa_n$ (as given by Eq. 334) with respect to $\lambda$ to zero to obtain the extremum values.

---

[19] The plane containing **n** is also characterized by being orthogonal to the tangent plane to the surface at $P$.

## 4.4 Principal Curvatures and Directions

Eq. 335 possesses two roots, $\lambda_1$ and $\lambda_2$, which according to the rules of polynomials are linked by the following relations:

$$\lambda_1 + \lambda_2 = \frac{gE - eG}{gF - fG} \qquad \lambda_1 \lambda_2 = \frac{fE - eF}{gF - fG} \qquad (336)$$

where $(gF - fG) \neq 0$. These roots represent the two directions corresponding to the two principal curvatures, $\kappa_1$ and $\kappa_2$, of the surface at the given point, as indicated above.

The following two vectors on the surface, which are defined in terms of the two $\lambda$ roots of the above quadratic equation, define the spatial directions corresponding to the principal curvatures:

$$\left(\frac{d\mathbf{r}}{du^1}\right)_1 = \frac{\partial \mathbf{r}}{\partial u^1} + \lambda_1 \frac{\partial \mathbf{r}}{\partial u^2} = \mathbf{E}_1 + \lambda_1 \mathbf{E}_2 \qquad (337)$$

$$\left(\frac{d\mathbf{r}}{du^1}\right)_2 = \frac{\partial \mathbf{r}}{\partial u^1} + \lambda_2 \frac{\partial \mathbf{r}}{\partial u^2} = \mathbf{E}_1 + \lambda_2 \mathbf{E}_2 \qquad (338)$$

These directions, which are called the principal directions or the curvature directions of the surface at point $P$, are orthogonal at non-umbilical points where $\kappa_1 \neq \kappa_2$. At umbilical points (see § 4.10), the normal curvature is the same in all directions and hence there are no principal directions to be orthogonal or every direction is a principal direction and hence there is no sensible meaning for being orthogonal. Consequently, at any point on a plane surface all directions are principal directions or, alternatively, there is no principal direction (depending on allowing more than two principal directions or not). Similarly, at any point on a sphere all directions are principal directions or there is no principal direction.

It is noteworthy that the principal directions are invariant with respect to permissible changes in surface representation and parameterization. However, we remark that although the principal directions in a given coordinate system of the ambient space are fixed and hence their position and orientation relative to the surface are invariant, their position and orientation with respect to the surface coordinates depend on the coordinate system employed to represent the surface.

The positions of the centers of curvature (see § 2.6) of the normal sections corresponding to the two principal curvatures at a given point $P$ on a surface $S$ are given in tensor notation by:

$$x_1^i = x_P^i + \frac{N_1^i}{|\kappa_1|} \qquad (339)$$

$$x_2^i = x_P^i + \frac{N_2^i}{|\kappa_2|} \qquad (340)$$

where $x_1^i$ and $x_2^i$ are the spatial coordinates of the first and second center of curvature corresponding to the two principal curvatures, $x_P^i$ are the spatial coordinates of $P$, $N_1^i$ and $N_2^i$ are the principal normal vectors of the two normal sections corresponding to the two principal curvatures, $\kappa_1$ and $\kappa_2$ are the principal curvatures of $S$ at $P$, and $i = 1, 2, 3$.

## 4.4 Principal Curvatures and Directions

We note that the principal normal vector **N** of a normal section at a given point $P$ on the surface is collinear with the unit normal vector **n** to the surface at $P$ and hence the principal normal vectors are in the same orientation for all normal sections at $P$. However, the two principal normal vectors corresponding to the two principal curvatures may be parallel or anti-parallel and hence we labeled them differently to be general.[20] We also remark that the two normal sections in the principal directions at $P$ are called the principal normal sections of the surface at $P$, while the centers of curvature of these principal normal sections are described as the principal centers of curvature of the surface at $P$.

According to one of the Euler theorems, the normal curvature $\kappa_n$ at a given point $P$ on a surface of class $C^2$ in a given direction can be expressed as a combination of the principal curvatures, $\kappa_1$ and $\kappa_2$, at $P$ as:

$$\kappa_n = \kappa_1 \cos^2 \theta + \kappa_2 \sin^2 \theta \tag{341}$$

where $\theta$ is the angle between the principal direction of $\kappa_1$ at $P$ and the given direction. Since the principal directions at non-umbilical points are orthogonal, $\theta$ could represent the angle with the other principal direction but with relabeling of the two kappas.

There are a number of invariant parameters of the surface at a given point $P$ on the surface which are defined in terms of the principal curvatures at $P$; these include:

1. The principal radii of curvature: $R_1 = \left|\frac{1}{\kappa_1}\right|$ and $R_2 = \left|\frac{1}{\kappa_2}\right|$.
2. The Gaussian curvature: $K = \kappa_1 \kappa_2$.
3. The mean curvature: $H = \frac{\kappa_1 + \kappa_2}{2}$.

Table 1 shows the restricting conditions on the principal curvatures, $\kappa_1$ and $\kappa_2$, for a number of common surfaces with simple geometric shapes (plane, cylinder, sphere, ellipsoid and hyperboloid of one sheet) and the effect on the Gaussian curvature $K$ and the mean curvature $H$.

The Gaussian curvature may also be called the "Riemannian curvature". However, the "Riemannian curvature" is usually used to label this type of curvature for general $n$D spaces while the "Gaussian curvature" is being used to label the special instance of it that applies to 2D spaces, and hence the Gaussian curvature is the Riemannian curvature of surfaces.

It should be remarked that some authors use "total curvature" for the "Gaussian curvature" and hence these two terms are synonym, while others use "total curvature" for the area integral $\int K d\sigma$ as used, for example, in the Gauss-Bonnet theorem (refer to § 4.8). In the present book, we use total curvature strictly for the integral and hence we label the Gaussian curvature with $K$ and the total curvature with $K_t$. Another remark is that some authors define $H$ as the sum of $\kappa_1$ and $\kappa_2$, that is: $H = \kappa_1 + \kappa_2$, rather than the average as defined above. In fact each one of these conventions has its merit. However, in the present book we define $H$ as the average, not the sum, of the two principal curvatures.

---

[20] Alternatively, we can use a single principal normal vector for both normal sections with the use of the principal curvatures instead of their absolute values. However, the sign of the second term on the right hand side of the above equations should be selected properly as it can be plus or minus depending on the choice of **n** direction.

## 4.4 Principal Curvatures and Directions

Table 1: The limiting conditions on the principal curvatures, $\kappa_1$ and $\kappa_2$, for a number of surfaces of simple geometric shapes alongside the corresponding mean curvature $H$ and Gaussian curvature $K$. Apart from the plane, the unit normal vector to the surface, **n**, is assumed to be in the outside direction.

|  | $\kappa_1$ | $\kappa_2$ | $H$ | $K$ |
|---|---|---|---|---|
| Plane | 0 | 0 | 0 | 0 |
| Cylinder | $\kappa_1 = 0$ | $\kappa_2 < 0$ | $H < 0$ | 0 |
| Sphere | $\kappa_1 = \kappa_2 < 0$ | $\kappa_2 = \kappa_1 < 0$ | $H < 0$ | $K > 0$ |
| Ellipsoid (Fig. 4) | $\kappa_1 < 0$ | $\kappa_2 < 0$ | $H < 0$ | $K > 0$ |
| Hyperboloid of one sheet (Fig. 5) | $\kappa_1 > 0$ | $\kappa_2 < 0$ | — | $K < 0$ |

In the neighborhood of a given point on a surface, the surface can be approximated by a quadratic expression involving the principal curvatures at that point. More formally, let $P$ be a point on a sufficiently smooth surface $S$ embedded in a 3D space coordinated by a rectangular Cartesian system $(x, y, z)$ with $P$ being above the origin, the tangent plane of $S(x, y)$ at $P$ being parallel to the $xy$ plane, and the principal directions being along the $x$ and $y$ coordinate lines. The equation of $S$ in the neighborhood of $P$ can then be expressed, up and including the quadratic terms, in the following form:

$$S(x,y) \simeq S(0,0) + \frac{\kappa_1 x^2}{2} + \frac{\kappa_2 y^2}{2} \qquad (342)$$

where $\kappa_1$ and $\kappa_2$ are the principal curvatures of $S$ at $P$. This means that in the immediate neighborhood of $P$, $S$ resembles a quadratic surface (see § 6.2) of the given form. The above form includes umbilical points (see § 4.10) where the principal directions can be arbitrarily chosen as the directions of the $x$ and $y$ coordinate lines.

The necessary and sufficient condition for a number $\kappa \in \mathbb{R}$ to be a principal curvature of a smooth surface $S$ at a given point $P$ and in a given direction $\frac{dv}{du}$, where $(du)^2 + (dv)^2 \neq 0$, is that the following equations are satisfied:

$$(e - \kappa E)du + (f - \kappa F)dv = 0 \qquad (343)$$
$$(f - \kappa F)du + (g - \kappa G)dv = 0 \qquad (344)$$

where $E, F, G, e, f, g$ are the coefficients of the first and second fundamental forms at $P$. It is worth noting that for simplicity in notation, we use $\kappa$ in these and the following equations to represent principal curvature. This use should not be confused with the curve curvature which is also symbolized by $\kappa$. However, for this case the curvature is equal (in magnitude at least) to the principal curvature since the latter is the curvature of a normal section and hence the use of $\kappa$ is justified.

The above equations can be cast in a matrix form as:

$$\begin{bmatrix} e - \kappa E & f - \kappa F \\ f - \kappa F & g - \kappa G \end{bmatrix} \begin{bmatrix} du \\ dv \end{bmatrix} = \begin{bmatrix} 0 \\ 0 \end{bmatrix} \qquad (345)$$

## 4.4 Principal Curvatures and Directions

This system of homogeneous linear equations has a non-trivial solution $(du, dv)$ *iff* the determinant of the coefficient matrix is zero, that is:

$$\begin{vmatrix} e - \kappa E & f - \kappa F \\ f - \kappa F & g - \kappa G \end{vmatrix} = \left(EG - F^2\right)\kappa^2 - \left(gE - 2fF + eG\right)\kappa + \left(eg - f^2\right) = 0 \quad (346)$$

Based on the given conditions, this quadratic equation in $\kappa$ has a non-negative discriminant and hence it possesses either two distinct real roots or a repeated real root. In the former case there are two distinct principal curvatures at $P$ corresponding to two orthogonal principal directions, while in the latter case the point is umbilical where all the normal sections at the point have the same "principal curvature" although there is no specific principal direction since each direction can be a principal direction. So in brief, a given real number $\kappa$ is a principal curvature of $S$ at $P$ *iff* it is a solution of Eq. 346.

From Eq. 346, it can be seen that the principal curvatures of a surface at a given point $P$ are the solutions of this quadratic equation and hence they are given by:

$$\kappa_{1,2} = \frac{gE - 2fF + eG \pm \sqrt{(gE - 2fF + eG)^2 - 4(EG - F^2)(eg - f^2)}}{2(EG - F^2)} \quad (347)$$

On dividing Eq. 346 by $a = EG - F^2$ (which is positive definite as established before) we obtain:

$$\kappa^2 - 2H\kappa + K = 0 \quad (348)$$

where $H$ and $K$ are the mean and Gaussian curvatures whose expressions in terms of the coefficients of the first and second fundamental forms are taken from Eqs. 383 and 356. Hence, Eq. 347 can be expressed compactly as:

$$\kappa_{1,2} = H \pm \sqrt{H^2 - K} \quad (349)$$

In fact, this formula can be obtained directly from Eq. 348 using the quadratic formula.

The above conditions about the principal curvatures may be stated rather differently in terms of the principal directions that is, for a non-umbilical point $P$ on a sufficiently smooth surface $S$, a direction $\frac{du^2}{du^1}$ is a principal direction of $S$ at $P$ *iff* the following condition is true:

$$(fE - eF)\,du^1 du^1 + (gE - eG)\,du^1 du^2 + (gF - fG)\,du^2 du^2 = 0 \quad (350)$$

The last equation, which is obtained from Eq. 335 by multiplying both sides with $(du^1)^2$, can be factored into two linear equations each of the form: $A\,du^1 + B\,du^2 = 0$ (with $A$ and $B$ being real constants) where these equations represent the two orthogonal principal directions.

Similarly, at a given non-umbilical point $P$ on a sufficiently smooth surface $S$, a direction $\frac{dv}{du}$ is a principal direction *iff* for a real number $\kappa$ the following relation holds true:

$$d\mathbf{n} = -\kappa\,d\mathbf{r} \quad (351)$$

## 4.4 Principal Curvatures and Directions

where:

$$d\mathbf{n} = \frac{\partial \mathbf{n}}{\partial u}du + \frac{\partial \mathbf{n}}{\partial v}dv \qquad d\mathbf{r} = \frac{\partial \mathbf{r}}{\partial u}du + \frac{\partial \mathbf{r}}{\partial v}dv \qquad (352)$$

If this condition is satisfied, then $\kappa$ is the principal curvature of $S$ at $P$ corresponding to the principal direction $\frac{dv}{du}$. Eq. 351 is known as the Rodrigues curvature formula. The obvious interpretation of the Rodrigues formula is that in any principal direction the two vectors $d\mathbf{n}$ and $d\mathbf{r}$ have the same orientation where the principal curvature $\kappa$ in that direction is the scale factor between the two vectors. From the Rodrigues curvature formula, the following subsidiary equations corresponding to the surface coordinate curves can be easily obtained:

$$\frac{\partial \mathbf{n}}{\partial u} = -\kappa \mathbf{E}_1 \qquad \frac{\partial \mathbf{n}}{\partial v} = -\kappa \mathbf{E}_2 \qquad (353)$$

On each non-umbilical point $P$ of a smooth surface $S$ an orthonormal moving "Darboux frame" can be defined. This frame consists of the vector triad $(\mathbf{d}_1, \mathbf{d}_2, \mathbf{n})$ where $\mathbf{d}_1$ and $\mathbf{d}_2$ are the unit vectors corresponding to the principal directions at $P$, and $\mathbf{n} = \mathbf{d}_1 \times \mathbf{d}_2$ is the unit normal vector to the surface at $P$. This is another moving frame in use in differential geometry in addition to the three previously-described frames: the $(\mathbf{T},\mathbf{N},\mathbf{B})$ frame, the $(\mathbf{E}_1, \mathbf{E}_2, \mathbf{n})$ frame and the $(\mathbf{n}, \mathbf{T}, \mathbf{u})$ frame (see § 1.4.5, 2.5, 3.2 and 4.1). The first of these frames, i.e. $(\mathbf{T},\mathbf{N},\mathbf{B})$, is associated with curves while the remaining three are associated with surfaces. What is common to all these four frames is that they are moving frames whose vectors can be used as basis sets for the embedding 3D space since each one of these sets consists of three linearly independent vectors. Also, all these sets, except $(\mathbf{E}_1, \mathbf{E}_2, \mathbf{n})$, are orthonormal.

When the $u$ and $v$ coordinate curves of a surface at a given point $P$ are aligned along the principal directions at $P$, the principal curvatures at $P$ will be given by (see § 4.2):

$$\kappa_1 = \frac{b_{11}}{a_{11}} = \frac{e}{E} \qquad \kappa_2 = \frac{b_{22}}{a_{22}} = \frac{g}{G} \qquad (354)$$

where the indexed $a$ and $b$ are the coefficients of the surface covariant metric and covariant curvature tensors, and $E, G, e, g$ are the coefficients of the first and second fundamental forms at $P$. This may be obtained from Eq. 316 where the last two terms in the sums will vanish for the $u$ coordinate curve since $dv = 0$ while the first two terms in the sums will vanish for the $v$ coordinate curve since $du = 0$. We note that we are assuming here a particular labeling of the $u$ and $v$ coordinate curves for the labeling of the two kappas to be appropriate, i.e. the $u$ coordinate curve is aligned along the first principal direction and the $v$ coordinate curve is aligned along the second principal direction.

It should be remarked that on an oriented and sufficiently smooth surface, the principal curvatures, $\kappa_1$ and $\kappa_2$, are continuous functions of the surface coordinates. Another remark is that the principal curvatures are the eigenvalues of the mixed type surface curvature tensor $b^\alpha_\beta$.

## 4.5 Gaussian Curvature

The Gaussian curvature, which may also be called the Riemannian curvature of the surface, represents a generalization of curve curvature to surfaces since it is the product of two curvatures of curves embedded in the surface and hence in this sense it is a 2D curvature. As given earlier, the Gaussian curvature $K$ at a given point $P$ on a surface is defined as the product of the two principal curvatures, $\kappa_1$ and $\kappa_2$, of the surface at $P$ that is:

$$K \equiv \kappa_1 \kappa_2 \qquad (355)$$

The Gaussian curvature of a surface at a given point $P$ on the surface is given by:

$$K = \frac{eg - f^2}{EG - F^2} = \frac{b}{a} = \frac{R_{1212}}{a} \qquad (356)$$

where $E, F, G, e, f, g$ are the coefficients of the first and second fundamental forms at $P$, $a$ and $b$ are the determinants of the surface covariant metric and covariant curvature tensors, and $R_{1212}$ is the component of the 2D covariant Riemann-Christoffel curvature tensor. From Eq. 231 we have:

$$R_{1212} = b_{11}b_{22} - b_{12}b_{21} = eg - f^2 = b \qquad (357)$$

where the indexed $b$ are the coefficients of the surface covariant curvature tensor, and hence the equalities in Eq. 356 are fully justified. As discussed previously (see § 1.4.10), the 2D Riemann-Christoffel curvature tensor has only one independent non-vanishing component which is represented by $R_{1212}$. Therefore, Eq. 356 provides a full link between the Gaussian curvature and the Riemann-Christoffel curvature tensor.

The above formulae (Eq. 356) are also based on the fact that the Gaussian curvature $K$ is the determinant of the mixed curvature tensor $b^\alpha_\beta$ of the surface, that is:

$$K = \det(b^\alpha_\beta) = \det(a^{\alpha\gamma} b_{\gamma\beta}) = \det(a^{\alpha\gamma})\det(b_{\gamma\beta}) = \frac{\det(b_{\gamma\beta})}{\det(a_{\alpha\gamma})} = \frac{b}{a} \qquad (358)$$

where the symbols are as defined previously. From Eq. 356, we see that the sign of $K$ (i.e. $K > 0$, $K < 0$ or $K = 0$) is the same as the sign of $b$ and the sign of $R_{1212}$ since $a$ is positive definite. We note that being the determinant of a tensor establishes the status of $K$ as an invariant under permissible coordinate transformations. We also note that the chain of formulae in Eq. 358 may be taken in the opposite direction starting primarily from $K = \frac{b}{a}$ or $K = \frac{R_{1212}}{a}$ as a definition or as a derived result from other arguments, and hence the statement $K = \det(b^\alpha_\beta)$ will be obtained as a secondary result.

Since both $R_{1212}$ (see Eq. 88) and $a$ depend exclusively on the surface metric tensor, Eq. 356 reveals that $K$ depends only on the first fundamental form coefficients and hence it is an intrinsic property of the surface (refer to § 4.7). The dependence of $K$ on the second fundamental form coefficients in Eq. 356 or Eq. 358 does not affect its qualification as an intrinsic property since this dependency is not indispensable as $K$ can be expressed in

## 4.5 Gaussian Curvature

terms of the first fundamental form coefficients exclusively. In fact, according to Eq. 262 even $b$ can be expressed exclusively in terms of the first fundamental form coefficients.

Because the Gaussian curvature is an invariant with respect to permissible coordinate transformations in 2D manifolds, we have:

$$K = \frac{R_{1212}}{a} = \frac{\bar{R}_{1212}}{\bar{a}} \tag{359}$$

where the barred and unbarred symbols represent the quantities in the barred and unbarred coordinate systems. The Gaussian curvature is also invariant with respect to the type of representation and parameterization of the surface. In particular, the Gaussian curvature is independent, in sign and magnitude, of the orientation of the surface which is based on the choice of the direction of the normal vector $\mathbf{n}$ to the surface. This is because a change in the direction of $\mathbf{n}$ will change the sign of the principal curvatures but not their absolute value and hence the magnitude is preserved. Furthermore, this change of sign will not affect the sign of the Gaussian curvature since both signs will be changed by the reversal of $\mathbf{n}$ direction and hence their product will not be affected. Therefore, the sign and magnitude of the Gaussian curvature are both preserved under this reversal.

From Table 1 we see that the Gaussian curvature of planes and cylinders are both identically zero. At the root of this is the fact that the Gaussian curvature is an intrinsic property and the cylinder is a developable surface obtained by wrapping a plane with no localized distortion by stretching or compression. Hence, the planes and cylinders possess identical first fundamental forms, as indicated previously in § 3.5, and consequently they have identical Gaussian curvature (also see § 4.7).

Since the magnitude of the normal curvature of a sphere of radius $R$ is $|\kappa_n| = \frac{1}{R}$ at any point on its surface and for any normal section in any direction, its Gaussian curvature is a constant given by $K = \frac{1}{R^2}$. For a Monge patch of the form $\mathbf{r}(u,v) = (u, v, f(u,v))$, the Gaussian curvature is given by:

$$K = \frac{f_{uu}f_{vv} - f_{uv}^2}{\left(1 + f_u^2 + f_v^2\right)^2} \tag{360}$$

where the subscripts $u$ and $v$ stand for partial derivatives of $f$ with respect to these surface coordinates. The last equation can be obtained by combining Eq. 356 (or Eq. 358) with Eqs. 201 and 226.

The Gaussian curvature of a surface of revolution generated by revolving a plane curve of class $C^2$ having the form $y = f(x)$ around the $x$-axis is given by:

$$K = -\frac{f_{xx}}{f\left(1 + f_x^2\right)^2} \tag{361}$$

where the subscript $x$ represents derivative of $f$ with respect to this variable.

At any point on a sufficiently smooth surface the Gaussian curvature satisfies the following relation:

$$\partial_u \mathbf{n} \times \partial_v \mathbf{n} = K\left(\mathbf{E}_1 \times \mathbf{E}_2\right) \tag{362}$$

## 4.5 Gaussian Curvature

On dot producting both sides with $\mathbf{n}$ we obtain:

$$\mathbf{n} \cdot (\partial_u \mathbf{n} \times \partial_v \mathbf{n}) = K\, \mathbf{n} \cdot (\mathbf{E}_1 \times \mathbf{E}_2) = K\sqrt{a} \tag{363}$$

where the last equality is based on Eq. 169. Hence:

$$K = \frac{\mathbf{n} \cdot (\partial_u \mathbf{n} \times \partial_v \mathbf{n})}{\sqrt{a}} \tag{364}$$

In the last equation the Gaussian curvature, which is an intrinsic property, is expressed in terms of the normal vector $\mathbf{n}$, which is an extrinsic entity, and its derivatives as well as the metric tensor.

There are surfaces with constant zero Gaussian curvature $K = 0$ (e.g. planes, cylinders and cones excluding the apex), surfaces with constant positive Gaussian curvature $K > 0$ (e.g. spheres with $K = \frac{1}{R^2}$ where $R$ is the sphere radius), and surfaces with constant negative Gaussian curvature $K < 0$ (e.g. Beltrami pseudo-spheres, seen in Fig. 14, with $K = -\frac{1}{\rho^2}$ where $\rho$ is the pseudo-radius of the pseudo-sphere). However, in general the Gaussian curvature is a variable function, in sign and magnitude, of the surface coordinates and hence a single surface can have Gaussian curvature of different signs and magnitudes. It is noteworthy that surfaces with constant non-zero Gaussian curvature $K$ may be described as spherical if $K > 0$ and pseudo-spherical if $K < 0$.

On scaling a surface up or down by a constant factor $c > 0$, the Gaussian curvature $K$ will scale by a factor of $\frac{1}{c^2}$. This is based on the fact that scaling the surface by a constant factor $c > 0$ is equivalent to scaling the coefficients of the surface metric tensor by $c^2$ (see Eq. 187) and scaling the coefficients of the surface curvature tensor by $c$ (see Eq. 214), and hence according to Eq. 356 or Eq. 358, $K$ will be scaled by a factor of $\frac{1}{c^2}$. This leads to the conclusion that when the surface curvature is a non-zero constant, the surface can be scaled up or down to make its Gaussian curvature 1 or $-1$ and hence simplify the formulations and calculations.

Based on the previous statements plus the fact that the Gaussian curvature is an intrinsic property, the Gaussian curvature is invariant with respect to all isometric transformations since, intrinsically, these transformations correspond to scaling the surface with unity even though the shape of the surface may have been deformed extrinsically. Hence, two isometric surfaces have identical Gaussian curvature at each pair of their corresponding points. However, two surfaces with equal Gaussian curvature at their corresponding points are not necessarily isometric. Yes, in the case of two sufficiently smooth surfaces with equal constant Gaussian curvature the two surfaces have local isometry. The details can be found in more advanced books on differential geometry.

In 3D manifolds, there is no compact surface of class $C^2$ with non-positive Gaussian curvature (i.e. $K \leq 0$) over the whole surface. Also, any compact surface, excluding the sphere, should have points with negative Gaussian curvature. In fact, the sphere is the only connected, compact and sufficiently smooth surface with constant Gaussian curvature. According to the Hilbert lemma, if $P$ is a point on a sufficiently smooth surface $S$ with $\kappa_1$ and $\kappa_2$ being the principal curvatures of $S$ at $P$ such that: $\kappa_1 > \kappa_2$, $\kappa_1$ is a

## 4.5 Gaussian Curvature

local maximum, and $\kappa_2$ is a local minimum, then the Gaussian curvature of $S$ at $P$ is non-positive, that is $K \leq 0$.

At a given point $P$ on a spherically-mapped (see § 3.10) and sufficiently smooth surface $S$, the ratio of the area of the spherical image $\bar{\mathfrak{R}}$ of a mapped region $\mathfrak{R}$ surrounding $P$ on $S$ to the area of $\mathfrak{R}$ converges to the absolute value of the Gaussian curvature at $P$ as $\mathfrak{R}$ shrinks to $P$ (see Eq. 303). From the Gauss-Bonnet theorem (see § 4.8), it can be shown that a surface will have identically-vanishing Gaussian curvature if at any point $P$ on the surface there are two families of geodesic curves (see § 5.7) in the neighborhood of $P$ intersecting at a constant angle.

From Eqs. 93 and 356, it can be seen that the Gaussian curvature $K$ of a sufficiently smooth surface represented by $\mathbf{r} = \mathbf{r}(u,v) = \mathbf{r}(u^1, u^2)$ can also be given by:

$$K = \frac{1}{a}\left[F_{uv} - \frac{1}{2}E_{vv} - \frac{1}{2}G_{uu} + a_{\alpha\beta}\left(\Gamma^{\alpha}_{12}\Gamma^{\beta}_{12} - \Gamma^{\alpha}_{11}\Gamma^{\beta}_{22}\right)\right] \qquad (\alpha,\beta=1,2) \qquad (365)$$

where $E, F, G$ are the coefficients of the first fundamental form, the subscripts $u$ and $v$ stand for partial derivatives with respect to these surface coordinates, $a = EG - F^2$ is the determinant of the surface covariant metric tensor and the indexed $a$ are its coefficients. The Christoffel symbols in the last equation are based on the surface metric.

The Gaussian curvature of a smooth surface of class $C^3$ represented by $\mathbf{r}(u,v)$ may also be given by:

$$K = \frac{1}{2\sqrt{a}}\left[\partial_u\left(\frac{FE_v}{E\sqrt{a}} - \frac{G_u}{\sqrt{a}}\right) + \partial_v\left(\frac{2F_u}{\sqrt{a}} - \frac{E_v}{\sqrt{a}} - \frac{FE_u}{E\sqrt{a}}\right)\right] \qquad (366)$$

where the symbols are as defined above. Accordingly, the Gaussian curvature of a surface of class $C^3$ represented by $\mathbf{r}(u,v)$ with orthogonal surface coordinate curves is given by:

$$K = -\frac{1}{2\sqrt{EG}}\left[\partial_u\left(\frac{G_u}{\sqrt{EG}}\right) + \partial_v\left(\frac{E_v}{\sqrt{EG}}\right)\right] \qquad (367)$$

This formula is obtained from the previous formula by setting $F = 0$ identically due to the orthogonality of the surface coordinate curves. The last formula will simplify to:

$$K = -\frac{\partial_{uu}\sqrt{G}}{\sqrt{G}} \qquad (368)$$

when the surface $\mathbf{r}(u,v)$ is represented by geodesic coordinates (see § 1.4.8) with the $u$ coordinate curves being geodesics and $u$ is a natural parameter.[21]

The Gaussian curvature $K$ can also be expressed in terms of the mean curvature $H$ (see § 4.6), that is:

$$K = (H+C)(H-C) = H^2 - C^2 \qquad (369)$$

---

[21] In brief, "geodesic coordinates" here stands for a coordinate system on a coordinate patch of a surface whose $u$ and $v$ coordinate curve families are orthogonal with one of these families ($u$ or $v$) being a family of geodesic curves.

## 4.5 Gaussian Curvature

where $C$ is given by:

$$C = \frac{\sqrt{(e^2G^2 + E^2g^2) - 4fF(eG + Eg) + 4(f^2EG + F^2eg) - 2egEG}}{2(EG - F^2)} \quad (370)$$

and $E, F, G, e, f, g$ are the coefficients of the first and second fundamental forms. This can be verified by transforming Eq. 369 to the following form: $C^2 = H^2 - K$ and substituting for $H$ and $K$ from Eqs. 383 and 356.

The Gaussian curvature of a surface $S$ at a given point $P$ on the surface is positive if all the surface points in a deleted neighborhood of $P$ on $S$ are on the same side of the tangent plane to $S$ at $P$. The Gaussian curvature is negative if for all deleted neighborhoods of $P$ on $S$ some points are on one side of the tangent plane and some are on the other side. The Gaussian curvature is zero if, in a deleted neighborhood, either all the points lie in the tangent plane or all the points are on one side except some which lie on a curve in the tangent plane. Hence:

1. A sphere has positive Gaussian curvature at all points.
2. A hyperbolic paraboloid (Fig. 8) has negative Gaussian curvature at all points. Similarly, the monkey saddle (Fig. 12) has negative Gaussian curvature at all points except the origin $(x, y, z) = (0, 0, 0)$ which is an umbilical point (see § 4.10) with zero Gaussian curvature.
3. A plane has zero Gaussian curvature at all points.
4. A cylinder has zero Gaussian curvature at all points.
5. A torus (Fig. 36) has points with positive Gaussian curvature (outer half), points with zero Gaussian curvature (top and bottom circles) and points with negative Gaussian curvature (inner half).

Based on the above statements, the Gaussian curvature of a developable surface (see § 6.4) is identically zero. Hence, beside the plane, there are other surfaces with constant zero Gaussian curvature such as cones, cylinders and tangent surfaces of space curves (refer to § 6.6).

Examples of the Gaussian curvature, $K$, for a number of simple surfaces are:

1. Plane: $K = 0$.
2. Sphere of radius $R$: $K = \frac{1}{R^2}$.
3. Torus parameterized by $x = (R + r\cos\phi)\cos\theta$, $y = (R + r\cos\phi)\sin\theta$ and $z = r\sin\phi$: $K = \frac{\cos\phi}{r(R + r\cos\phi)}$.

The total curvature $K_t$ is defined as the area integral of the Gaussian curvature $K$ over a surface or a patch of a surface, $S$, that is:

$$K_t = \iint_S K \, d\sigma \quad (371)$$

where $d\sigma$ symbolizes infinitesimal area element on the surface and where $K$ is a function of the surface coordinates in general.

From Eqs. 206 and 282, it can be seen that the total curvature $K_t$ may be given by:

$$K_t \equiv \iint_S K \, d\sigma = \iint_S K \, |\mathbf{E}_1 \times \mathbf{E}_2| \, du\, dv = \iint_S \operatorname{sgn}(K) \, |\partial_u \mathbf{n} \times \partial_v \mathbf{n}| \, du\, dv \quad (372)$$

## 4.5 Gaussian Curvature

Figure 36: Points of torus with positive Gaussian curvature (outer blue), points with zero Gaussian curvature (middle yellow) and points with negative Gaussian curvature (inner red).

where sgn $(K)$ is the sign function of $K$ as a function of the surface coordinates, $u$ and $v$.

The Riemann-Christoffel curvature tensor is related to the Gaussian curvature through the absolute permutation tensor of the surface by the following relation:

$$R_{\alpha\beta\gamma\delta} = K \underline{\epsilon}_{\alpha\beta} \underline{\epsilon}_{\gamma\delta} \tag{373}$$

where the indexed $\underline{\epsilon}$ are the 2D covariant absolute permutation tensors and all the indices range over 1 and 2. On multiplying both sides of the last equation by $\underline{\epsilon}^{\alpha\beta} \underline{\epsilon}^{\gamma\delta}$ we get:

$$\underline{\epsilon}^{\alpha\beta} \underline{\epsilon}^{\gamma\delta} R_{\alpha\beta\gamma\delta} = K \underline{\epsilon}^{\alpha\beta} \underline{\epsilon}^{\gamma\delta} \underline{\epsilon}_{\alpha\beta} \underline{\epsilon}_{\gamma\delta} \tag{374}$$

Now, since $\underline{\epsilon}^{\alpha\beta} \underline{\epsilon}_{\alpha\beta} = \underline{\epsilon}^{\gamma\delta} \underline{\epsilon}_{\gamma\delta} = 2$, the last equation becomes:

$$K = \frac{1}{4} \underline{\epsilon}^{\alpha\beta} \underline{\epsilon}^{\gamma\delta} R_{\alpha\beta\gamma\delta} = \frac{1}{4} \underline{\epsilon}^{\alpha\beta} \underline{\epsilon}^{\gamma\delta} \left( b_{\alpha\gamma} b_{\beta\delta} - b_{\alpha\delta} b_{\beta\gamma} \right) \tag{375}$$

where the indexed $b$ are the components of the surface covariant curvature tensor, and where the last step is based on Eq. 227. The last equation is a demonstration of the fact that $K$ is an absolute rank-0 tensor since it is represented in both equalities by a combination of absolute tensors with all the indices of these tensors being consumed by contraction.

The Gaussian curvature is also linked to the Riemann-Christoffel curvature tensor, through the surface metric tensor, by the following relation:

$$R_{\alpha\beta\gamma\delta} = K \left( a_{\alpha\gamma} a_{\beta\delta} - a_{\alpha\delta} a_{\beta\gamma} \right) \tag{376}$$

## 4.5 Gaussian Curvature

In fact, Eq. 356 is an instance of the last equation with $\alpha = \gamma = 1$ and $\beta = \delta = 2$. The other combinations of index values provide the link between $K$ and the other elements of $R_{\alpha\beta\gamma\delta}$. We note that Eq. 376 may be extended to $n$D spaces for $n > 2$ and with $K$ (representing Riemannian curvature) being constant but this is out of the scope of this book.

The Gaussian curvature $K$ may also be given by the following relation:

$$K = \frac{1}{2}\underline{\epsilon}^{\alpha\beta}\underline{\epsilon}^{\gamma\delta}b_{\gamma\alpha}b_{\delta\beta} \qquad (377)$$

where the indexed $\underline{\epsilon}$ are the 2D contravariant absolute permutation tensors. From Eqs. 227 and 373, it can be seen that the Gaussian curvature and the surface curvature tensor are also related by:

$$K\underline{\epsilon}_{\alpha\beta}\underline{\epsilon}_{\gamma\delta} = b_{\alpha\gamma}b_{\beta\delta} - b_{\alpha\delta}b_{\beta\gamma} \qquad (378)$$

Other formulae for the Gaussian curvature (in terms of the surface basis vectors, their derivatives and the coefficients of the first fundamental form) may also be obtained from the formula $K = \frac{b}{a}$ by manipulating $b$ as follows:

$$\begin{aligned} b &= eg - f^2 & (379) \\ &= \frac{(\partial_u \mathbf{E}_1 \cdot \mathbf{E}_1 \times \mathbf{E}_2)(\partial_v \mathbf{E}_2 \cdot \mathbf{E}_1 \times \mathbf{E}_2) - (\partial_v \mathbf{E}_1 \cdot \mathbf{E}_1 \times \mathbf{E}_2)^2}{a} \\ &= \frac{(\partial_u \mathbf{E}_1 \cdot \mathbf{E}_1 \times \mathbf{E}_2)(\partial_v \mathbf{E}_2 \cdot \mathbf{E}_1 \times \mathbf{E}_2) - (\partial_v \mathbf{E}_1 \cdot \mathbf{E}_1 \times \mathbf{E}_2)^2}{EG - F^2} \\ &= \frac{(\partial_u \mathbf{E}_1 \cdot \mathbf{E}_1 \times \mathbf{E}_2)(\partial_v \mathbf{E}_2 \cdot \mathbf{E}_1 \times \mathbf{E}_2) - (\partial_v \mathbf{E}_1 \cdot \mathbf{E}_1 \times \mathbf{E}_2)^2}{|\mathbf{E}_1 \times \mathbf{E}_2|^2} \end{aligned}$$

where these steps are based on Eqs. 253-255 and Eq. 170 as well as the obvious fact that $a = EG - F^2$. Hence:

$$\begin{aligned} K &= \frac{b}{a} & (380) \\ &= \frac{(\partial_u \mathbf{E}_1 \cdot \mathbf{E}_1 \times \mathbf{E}_2)(\partial_v \mathbf{E}_2 \cdot \mathbf{E}_1 \times \mathbf{E}_2) - (\partial_v \mathbf{E}_1 \cdot \mathbf{E}_1 \times \mathbf{E}_2)^2}{a^2} \\ &= \frac{(\partial_u \mathbf{E}_1 \cdot \mathbf{E}_1 \times \mathbf{E}_2)(\partial_v \mathbf{E}_2 \cdot \mathbf{E}_1 \times \mathbf{E}_2) - (\partial_v \mathbf{E}_1 \cdot \mathbf{E}_1 \times \mathbf{E}_2)^2}{(EG - F^2)^2} \\ &= \frac{(\partial_u \mathbf{E}_1 \cdot \mathbf{E}_1 \times \mathbf{E}_2)(\partial_v \mathbf{E}_2 \cdot \mathbf{E}_1 \times \mathbf{E}_2) - (\partial_v \mathbf{E}_1 \cdot \mathbf{E}_1 \times \mathbf{E}_2)^2}{|\mathbf{E}_1 \times \mathbf{E}_2|^4} \end{aligned}$$

Finally, on a 2D surface, the Gaussian curvature $K$ is related to the Ricci curvature scalar $\mathcal{R}$ (see § 1.4.11) by the following relation:

$$|K| = \frac{|\mathcal{R}|}{2} \qquad (381)$$

As seen, the Gaussian curvature is just a constant multiple of the Ricci curvature scalar of the surface and hence they are essentially the same. Details about the signs of these curvature parameters should be sought in more expanded textbooks on differential geometry and tensor analysis.

## 4.6 Mean Curvature

The mean curvature of a surface at a given point $P$ is a measure of the rate of change of area of the surface elements in the neighborhood of $P$ with respect to the surface coordinates. As given earlier, the mean curvature $H$ is defined as the average[22] of the two principal curvatures, $\kappa_1$ and $\kappa_2$, that is:

$$H \equiv \frac{\kappa_1 + \kappa_2}{2} \tag{382}$$

The mean curvature is given by the following formula:

$$H = \frac{eG - 2fF + gE}{2(EG - F^2)} = \frac{\operatorname{tr}(b^\alpha_\beta)}{2} = \frac{b^\alpha_\alpha}{2} \qquad (\alpha, \beta = 1, 2) \tag{383}$$

where $E, F, G, e, f, g$ are the coefficients of the first and second fundamental forms, the indexed $b$ represent the surface mixed curvature tensor, and tr stands for the trace of matrix. The first equality can be obtained by combining Eq. 382 with Eq. 347, while the second equality can be verified by taking the trace of $b^\alpha_\beta$ as given by Eq. 223. The third equality is just a matter of different symbolism according to the matrix and tensor notations.

Unlike the Gaussian curvature, the sign of the mean curvature $H$ is dependent on the choice of the direction of the unit normal vector to the surface, $\mathbf{n}$. This can be seen from Eq. 382 where the signs of both kappas will be reversed by the change of $\mathbf{n}$ direction although the magnitude of kappas, and hence the magnitude of $H$, will not be affected by this change. Like the Gaussian curvature, the mean curvature is invariant under permissible coordinate transformations and representations as long as the surface orientation is preserved. Being half the trace of a tensor establishes the status of $H$ as an invariant under permissible coordinate transformations.

Examples of the mean curvature, $H$, for a number of simple surfaces are:
1. Plane: $H = 0$.
2. Sphere of radius $R$: $|H| = \frac{1}{R}$. As stated above, the sign of $H$ depends on the choice of $\mathbf{n}$ direction being inward or outward.
3. Torus parameterized by $x = (R + r\cos\phi)\cos\theta$, $y = (R + r\cos\phi)\sin\theta$ and $z = r\sin\phi$: $|H| = \left|\frac{R + 2r\cos\phi}{2r(R + r\cos\phi)}\right|$. The sign of $H$ in this case depends on the location of the point on the surface as well as the choice of $\mathbf{n}$ direction.

For a Monge patch of the form $\mathbf{r}(u, v) = (u, v, f(u, v))$, the mean curvature is given by:

$$H = \frac{(1 + f_v^2) f_{uu} - 2 f_u f_v f_{uv} + (1 + f_u^2) f_{vv}}{2 (1 + f_u^2 + f_v^2)^{3/2}} \tag{384}$$

---

[22] Or the sum depending on the authors although it will not be a mean anymore.

where the subscripts $u$ and $v$ stand for partial derivatives of $f$ with respect to these surface coordinates. This equation can be obtained from the first equality of Eq. 383 where the coefficients of the first and second fundamental forms of Monge patch are obtained from Eqs. 201 and 226.

The mean curvature may be considered as the 2D equivalent of the geodesic curvature in 1D. The equivalence can be understood in the sense that the mean curvature is a measure for extremizing surface area while the geodesic curvature is a measure for extremizing curve length. Accordingly, the 2D minimal surfaces (see § 6.7) correspond to the 1D geodesic curves (see § 5.7).

## 4.7 Theorema Egregium

The essence of Gauss *Theorema Egregium* or *Remarkable Theorem* is that the Gaussian curvature $K$ of a surface is an intrinsic property of the surface and hence it can be expressed as a function of the coefficients of the first fundamental form and their partial derivatives alone with no involvement of the coefficients of the second fundamental form. This can be guessed for example from the last part of Eq. 356. In fact, even the first part of Eq. 356 can be used in this argument since $b$ can be expressed purely in terms of the coefficients of the first fundamental form and their derivatives according to Eq. 262.

The essence of *Theorema Egregium*, as a statement of the fact that certain types of curvature are intrinsic to the surface, is contained in several forms and equations; some of which are indicated in this book when they occur. For example, Eq. 227 which links the surface curvature tensor to the Riemann-Christoffel curvature tensor (which is an intrinsic property of the surface and is related to the Gaussian curvature by Eq. 373 for instance) can be regarded as a statement of *Theorema Egregium* since it expresses a form of surface curvature represented by a certain combination of the coefficients of the curvature tensor in terms of a combination of purely intrinsic surface parameters.

An example may be given to demonstrate the significance of *Theorema Egregium* that is, if a piece of plane is rolled into a cylinder of radius $R$, then $\kappa_1, \kappa_2, H$ will change from $0, 0, 0$ to $\frac{1}{R}, 0, \frac{1}{2R}$ where, for the cylinder, we are assuming a normal unit vector $\mathbf{n}$ in the inner direction. However, as a consequence of *Theorema Egregium*, $K$ will not change since $K$ is dependent exclusively on the first fundamental form which is the same for planes and cylinders as stated previously.

According to *Theorema Egregium*, the Gaussian curvature of a sufficiently smooth surface of class $C^3$ at a given point $P$ can be represented by the following function of the coefficients of the first fundamental form and their partial derivatives at $P$:

$$K = \frac{1}{(EG-F^2)^2} \left\{ \begin{vmatrix} C & F_v - \frac{1}{2}G_u & \frac{1}{2}G_v \\ \frac{1}{2}E_u & E & F \\ F_u - \frac{1}{2}E_v & F & G \end{vmatrix} - \begin{vmatrix} 0 & \frac{1}{2}E_v & \frac{1}{2}G_u \\ \frac{1}{2}E_v & E & F \\ \frac{1}{2}G_u & F & G \end{vmatrix} \right\} \quad (385)$$

$$= \frac{1}{(EG-F^2)^2} \left\{ \begin{vmatrix} C & [22,1] & [22,2] \\ [11,1] & a_{11} & a_{12} \\ [11,2] & a_{21} & a_{22} \end{vmatrix} - \begin{vmatrix} 0 & [21,1] & [21,2] \\ [21,1] & a_{11} & a_{12} \\ [21,2] & a_{21} & a_{22} \end{vmatrix} \right\}$$

where $C = \frac{1}{2}(-E_{vv} + 2F_{uv} - G_{uu})$ and the subscripts $u$ and $v$ stand for partial derivatives with respect to these surface coordinates. The other symbols and notations are as defined previously.

## 4.8 Gauss-Bonnet Theorem

This theorem ties the geometry of surfaces to their topology. There are several variants of this theorem; some of which are local while others are global. Due to the importance and subtlety of this theorem we give two variants of the theorem and several examples from both plane and twisted surfaces. According to the Gauss-Bonnet theorem, if $\mathfrak{S}$ is a simply connected region on a surface of class $C^3$ where $\mathfrak{S}$ is bordered by a finite number $m$ of piecewise regular curves $C_j$ that meet in $n$ corners then we have:

$$\sum_{j=1}^{m} \int_{C_j} \kappa_g + \sum_{k=1}^{n} \phi_k + \iint_{\mathfrak{S}} K d\sigma = 2\pi \tag{386}$$

where the first sum is over the curves while the second sum is over the corners, $\kappa_g$ is the geodesic curvature of the curves $C_j$ as a function of their coordinates, $\phi_k$ are the exterior angles of the corners and $K$ is the Gaussian curvature of $\mathfrak{S}$ as a function of the coordinates over $\mathfrak{S}$. The geodesic and Gaussian curvatures in the above formulation should be continuous and finite over their domain. As indicated previously, the term $\iint_{\mathfrak{S}} K d\sigma$, which represents the area integral of the Gaussian curvature over the region $\mathfrak{S}$ of the above-described surface, is called the total curvature $K_t$ of $\mathfrak{S}$.

We note that the corners indicated in the last paragraph can be defined as the points of discontinuity of the tangents of the boundary curves. The angles of these corners are therefore defined as the angles between the tangent vectors at the points of discontinuity when traversing the boundary curves in a predefined sense. As indicated above, these angles are exterior to the region surrounded by the curves. Sometimes, "artificial corners" at regular points are introduced for convenience to establish an argument; in which case the exterior angle is zero. Several examples related to artificial corners will be given in the forthcoming parts of the book.

It should be remarked that the form of the Gauss-Bonnet theorem given by Eq. 386 may be labeled as a local variant of the theorem although its locality may not be obvious. However, it can be justified by its application in principle to a part of the surface in comparison to the forthcoming global variant of the theorem which applies to the whole surface and involves the Euler characteristic and the genus of the surface which are global features of the surface. Anyway, these labels are not of crucial importance as long as the theorem and its significance are understood and appreciated.

Some examples for the application of the above form of the Gauss-Bonnet theorem are given below:

1. A disc in a plane with radius $R$ where Eq. 386 becomes:

$$\frac{1}{R} 2\pi R + 0 + 0 = 2\pi + 0 + 0 \equiv 2\pi \tag{387}$$

## 4.8 Gauss-Bonnet Theorem

which is an identity.

2. A semi-circular disc in a plane with radius $R$ where Eq. 386 becomes:

$$\left(\frac{1}{R}\pi R + 0 \times 2R\right) + 2\left(\frac{\pi}{2}\right) + 0 = \pi + \pi + 0 \equiv 2\pi \tag{388}$$

which is an identity again.

3. A spherical triangle (Fig. 37) on a sphere of radius $R$ whose sides are two half meridians connecting a pole to the equator and one quarter of an equatorial parallel and all of its three corners are right angles where Eq. 386 becomes:

$$\left[0\left(3 \times \frac{\pi R}{2}\right)\right] + 3\left(\frac{\pi}{2}\right) + \frac{1}{R^2}\frac{4\pi R^2}{8} = 0 + \frac{3\pi}{2} + \frac{\pi}{2} \equiv 2\pi \tag{389}$$

4. The upper half of a sphere (or a hemisphere in general) of radius $R$ where Eq. 386 becomes:

$$0\left(2\pi R\right) + 0 + \frac{1}{R^2}2\pi R^2 = 0 + 0 + 2\pi \equiv 2\pi \tag{390}$$

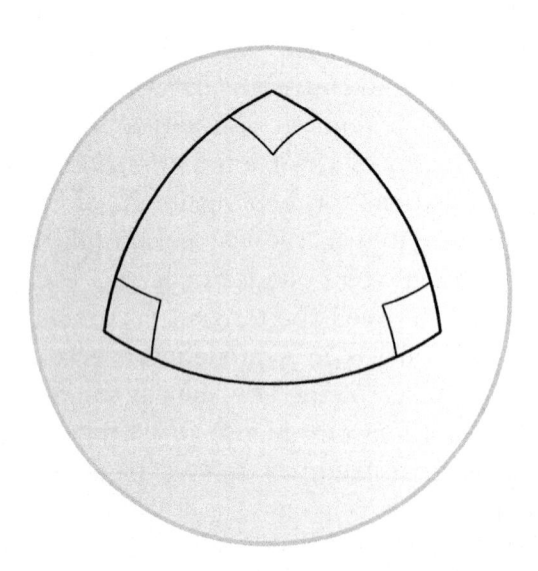

Figure 37: A spherical triangle with three right angles on the surface of a sphere. The three sides of this spherical triangle are arcs of great circles.

The fact that the sum of the interior angles of a planar triangle is equal to $\pi$ can also be regarded as an instance of the Gauss-Bonnet theorem since for a planar triangle Eq. 386 becomes:

$$0 + \sum_{i=1}^{3}(\pi - \theta_i) + 0 = 3\pi - \sum_{i=1}^{3}\theta_i = 2\pi \tag{391}$$

where $\theta_i$ are the interior angles of the triangle and hence $\sum_{i=1}^{3}\theta_i = \pi$ as it should be.

## 4.8 Gauss-Bonnet Theorem

By a similar argument, we can obtain the sum of the interior angles of a planar polygon of $n$ sides ($n > 2$) using the Gauss-Bonnet theorem, that is:

$$0 + \sum_{i=1}^{n}(\pi - \theta_i) + 0 = n\pi - \sum_{i=1}^{n}\theta_i = 2\pi \tag{392}$$

where $\theta_i$ are the interior angles of the $n$-polygon and hence $\sum_{i=1}^{n}\theta_i = (n-2)\pi$ as it should be.

The fact that the perimeter of a planar circle of radius $R$ is $2\pi R$ can be regarded as another instance of the Gauss-Bonnet theorem since for such a planar circle Eq. 386 becomes:

$$\frac{1}{R}L + 0 + 0 = 2\pi \tag{393}$$

where $L$ is the length of the circle perimeter and hence $L = 2\pi R$ which is the required result.

As a result of the Gauss-Bonnet theorem, the sum $\theta_s$ of the interior angles of a geodesic triangle on a surface with Gaussian curvature of constant sign is:

1. $\theta_s < \pi$ iff $K < 0$.
2. $\theta_s = \pi$ iff $K = 0$.
3. $\theta_s > \pi$ iff $K > 0$.

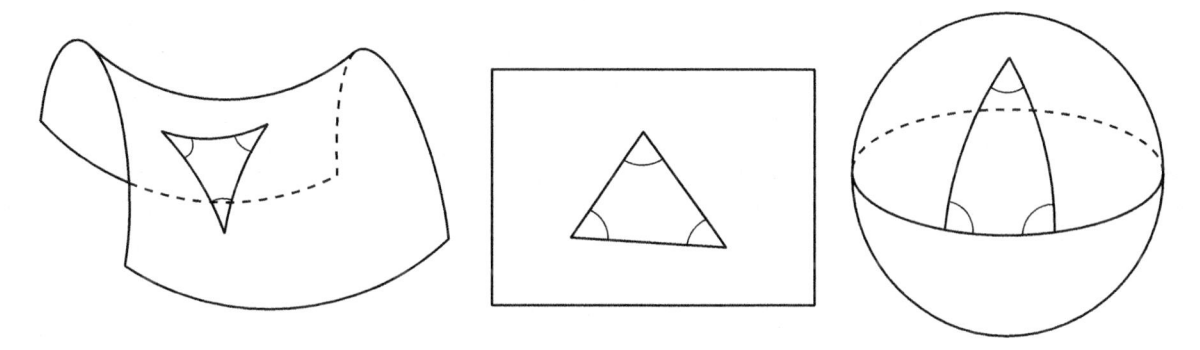

Figure 38: Geodesic triangles on a surface with negative Gaussian curvature (left frame), a surface with zero Gaussian curvature (middle frame), and a surface with positive Gaussian curvature (right frame).

These cases are depicted in Fig. 38. This shows that the total curvature provides the excess over $\pi$ for the sum when $K > 0$ on the surface and the deficit when $K < 0$. The vanishing total curvature in the case of $K = 0$ is the intermediate case where the total curvature term has no contribution to the sum. This can be seen from Eq. 386 which, for a geodesic triangle, will reduce to:

$$0 + (3\pi - \theta_s) + \iint_{\mathfrak{S}} K d\sigma = 2\pi \quad \implies \quad \theta_s = \pi + \iint_{\mathfrak{S}} K d\sigma \tag{394}$$

We remark that "geodesic triangle" is a triangle with geodesic sides and hence $\kappa_g = 0$ identically over its boundary (see § 5.7). Also, "triangle" here and in the spherical triangle

## 4.8 Gauss-Bonnet Theorem

example related to Fig. 37 is more general than a three-side planar polygon with three straight segments as it can be on a curved surface with curved non-planar sides.

As a consequence of the findings in the last paragraph, two geodesic curves on a simply connected patch of a surface with negative Gaussian curvature cannot intersect at two points because on introducing a vertex at a regular point on one curve we will have an artificial corner with zero exterior angle and hence $\pi$ interior angle. We will then have a geodesic triangle with $\theta_s > \pi$ on a surface over which $K < 0$, in violation of the above-stated condition. By a similar argument to the argument in the previous paragraph, the area of a geodesic polygon on a surface with constant non-zero Gaussian curvature is determined by the sum of the polygon interior angles $\theta_s$. This can also be seen from Eq. 386 which in this case will be reduced to:

$$0 + (n\pi - \theta_s) + K \iint_{\mathfrak{S}} d\sigma = 2\pi \quad \Longrightarrow \quad \iint_{\mathfrak{S}} d\sigma = \frac{\theta_s + (2-n)\pi}{K} \tag{395}$$

where $n > 2$ is the number of sides of the polygon. As for geodesic triangle, "geodesic polygon" is a polygon with geodesic sides and hence $\kappa_g = 0$ identically over its boundary. Again, "polygon" here is general and hence it includes curvilinear polygon on curved surface with curved non-planar sides.

It is worth noting that because the geodesic curvature is an intrinsic property, as discussed in § 4.3, the Gauss-Bonnet theorem, as given by Eq. 386, is another indication to the fact that the Gaussian curvature (as well as the total curvature) is an intrinsic property and hence it is another demonstration of *Theorema Egregium* (see § 4.7).

The Gauss-Bonnet theorem has also a global variant which links the Euler characteristic $\chi$, which is a topological invariant of the surface, to the Gaussian curvature $K$, which is a geometric invariant of the surface. This global form of the Gauss-Bonnet theorem states that: on a compact orientable surface $S$ of class $C^3$ these two invariants are linked through the following equation:

$$\iint_S K d\sigma = 2\pi\chi \tag{396}$$

Now, since $\chi$ is a topological invariant of the surface, Eq. 396 reveals that the total curvature is also a topological invariant of the surface.

The global Gauss-Bonnet theorem can be used to determine the total curvature $K_t$ of a surface. For example, the Euler characteristic of a sphere is 2 and hence from Eq. 396 it can be concluded that its total curvature is $K_t = 4\pi$ with no need for evaluating the area integral. Similarly, the Euler characteristic of a torus is 0 and hence it can be concluded immediately that its total curvature is $K_t = 0$ with no need for evaluating the integral. The Euler characteristics of the sphere and torus in these examples can be obtained easily by polygonal decomposition, as described in § 1.4.1. For example, the Euler characteristic of the sphere can be calculated by dividing the surface of the sphere to 4 curved polygonal faces with 4 vertices and 6 edges and hence the Euler characteristic is:

$$\chi = \mathcal{V} + \mathcal{F} - \mathcal{E} = 4 + 4 - 6 = 2 \tag{397}$$

as seen in Fig. 16.

The global Gauss-Bonnet theorem can also be used in the opposite direction, that is it may be used for determining the Euler characteristic of a surface knowing its Gaussian, and hence total, curvature although in most cases this may be of little use practically. For instance, the Gaussian curvature of a sphere of radius $R$ is $\frac{1}{R^2}$ at every point on the sphere and hence its total curvature is $K_t = K\sigma = \frac{1}{R^2} 4\pi R^2 = 4\pi$, therefore from Eq. 396 its Euler characteristic is $\chi = \frac{4\pi}{2\pi} = 2$.

The Gauss-Bonnet theorem can also be used to find the total curvature of a smooth surface which is topologically-equivalent (i.e. homeomorphic) to another surface with known total curvature without need for any calculation. For example, the ellipsoid (Fig. 4) is homeomorphic to the sphere and hence they have the same Euler characteristic. Therefore, according to Eq. 396 they have the same total curvature which is $4\pi$ as known from the aforementioned sphere example. As a consequence, the total curvature $K_t$ of a smooth surface with a complex shape can be obtained from the Gauss-Bonnet theorem by reducing the surface to a topologically-equivalent simpler surface whose total curvature can be evaluated promptly.

As seen before (refer to § 1.4.1), for an orientable surface of genus $\mathfrak{g}$ the Euler characteristic is given by: $\chi = 2(1 - \mathfrak{g})$, and hence its total curvature is given by:

$$K_t = 2\pi\chi = 4\pi(1 - \mathfrak{g}) \tag{398}$$

So, for a compact orientable complexly-shaped surface which can be reduced to a sphere with 2 handles the total curvature is $K_t = 4\pi(1 - 2) = -4\pi$. Similarly, the genus of a torus is $\mathfrak{g} = 1$ and hence its total curvature is $K_t = 4\pi(1 - 1) = 0$, as found earlier by another method. Hence, the total curvature of any complexly-shaped surface that can be reduced to a torus is zero.

The important and obvious implication of the global variant of the Gauss-Bonnet theorem that can be concluded from the previous discussion is that the total curvature of a closed surface is dependent on its genus and Euler characteristic and not on its geometric shape and hence it is a topological parameter of the surface as stated before.

## 4.9 Local Shape of Surface

Using the principal curvatures, $\kappa_1$ and $\kappa_2$, a point $P$ on a surface is classified according to the shape of the surface in the close proximity of $P$ as:
1. Flat when $\kappa_1 = \kappa_2 = 0$, and hence $K = H = 0$ (see Eqs. 355 and 382).
2. Parabolic when either $\kappa_1 = 0$ and $\kappa_2 \neq 0$ or $\kappa_2 = 0$ and $\kappa_1 \neq 0$, and hence $K = 0$ and $H \neq 0$.
3. Elliptic when either $\kappa_1 > 0$ and $\kappa_2 > 0$ or $\kappa_1 < 0$ and $\kappa_2 < 0$, and hence $K > 0$.
4. Hyperbolic when $\kappa_1 > 0$ and $\kappa_2 < 0$, and hence $K < 0$.

These constraints on $\kappa_1$ and $\kappa_2$, and hence on $K$ and $H$, are sufficient and necessary conditions for determining the type of the surface point as described above.

The following are some examples for the above classification:
1. The points of plane are flat.
2. The points of cone (excluding the apex) and the points of cylinder are parabolic.

## 4.9 Local Shape of Surface

3. The points of ellipsoid (Fig. 4) are elliptic.
4. The points of catenoid (Fig. 10) are hyperbolic.

Surfaces normally contain points of different shapes. For example, the torus has elliptic points on its outside half, parabolic points on its top and bottom parallels,[23] and hyperbolic points on its inside half, as seen in Fig. 36. However, there are some types of surface whose all points are of the same shape; e.g. all points of planes are flat, all points of spheres are elliptic, all points of catenoids are hyperbolic, and all points of cylinders are parabolic.

The above classification regarding the local shape can also be based on the determinant $b$ of the covariant curvature tensor and the coefficients $e, f, g$ of the second fundamental form of the surface where:

1. $b = eg - f^2 = 0$ and $e = f = g = 0$ for flat points.
2. $b = eg - f^2 = 0$ and $e^2 + f^2 + g^2 \neq 0$ for parabolic points.
3. $b = eg - f^2 > 0$ for elliptic points.
4. $b = eg - f^2 < 0$ for hyperbolic points.

Considering Eqs. 356 and 383 plus the fact that the first fundamental form is positive definite and hence $a > 0$, this classification which is based on $b$ and $e, f, g$ is equivalent to the previous classification which is based on $K$ and $H$.

The above classification of the shape of a surface in the immediate neighborhood of a point (i.e. being flat, parabolic, elliptic or hyperbolic) is an invariant property with respect to permissible coordinate transformations. This can be concluded from the dependence of the classification on the sign of $b$ as explained above, plus Eq. 221 where the square of the Jacobian (which is real) is positive and hence the sign of $b$ and $\bar{b}$ is the same. The classification is also independent of the representation and parameterization of the surface since these point types are real geometric properties of the surface in their local definitions.

The invariance of the shape type of the surface points, as explained in the previous statements, holds true even for the transformations that reverse the direction of the normal vector to the surface, **n**, because the classification depends on the Gaussian curvature which is invariant even under this type of transformations (refer to § 4.5). Regarding the distinction between the flat and parabolic points which involves $H$ as well, the distinction is not affected since it depends on the magnitude of $H$ (i.e. being zero or not) and not on its sign and the magnitude is not affected by such transformations.

In the immediate neighborhood of an elliptic point $P$ of a surface $S$, the surface lies completely on one side of the tangent plane to $S$ at $P$ (Fig. 39 a), while at a hyperbolic point the tangent plane cuts through $S$ and hence some parts of $S$ are on one side of the tangent plane while other parts are on the other side (Fig. 39 b). In the neighborhood of a parabolic point, the surface lies entirely on one side of the tangent plane except for some points on a curve which lies in the tangent plane itself (Fig. 39 c).[24] As for planar points, the neighborhood of the point lies in the tangent plane.

---

[23] These parallels correspond to the two circles contacting its two tangent planes at the top and bottom which are perpendicular to its axis of symmetry (see Fig. 36).

[24] This is the common case, however in some exceptional cases the surface in the neighborhood of a parabolic point lies on both sides of the tangent plane.

4.9 Local Shape of Surface

(a) Elliptic point

(b) Hyperbolic point

(c) Parabolic point

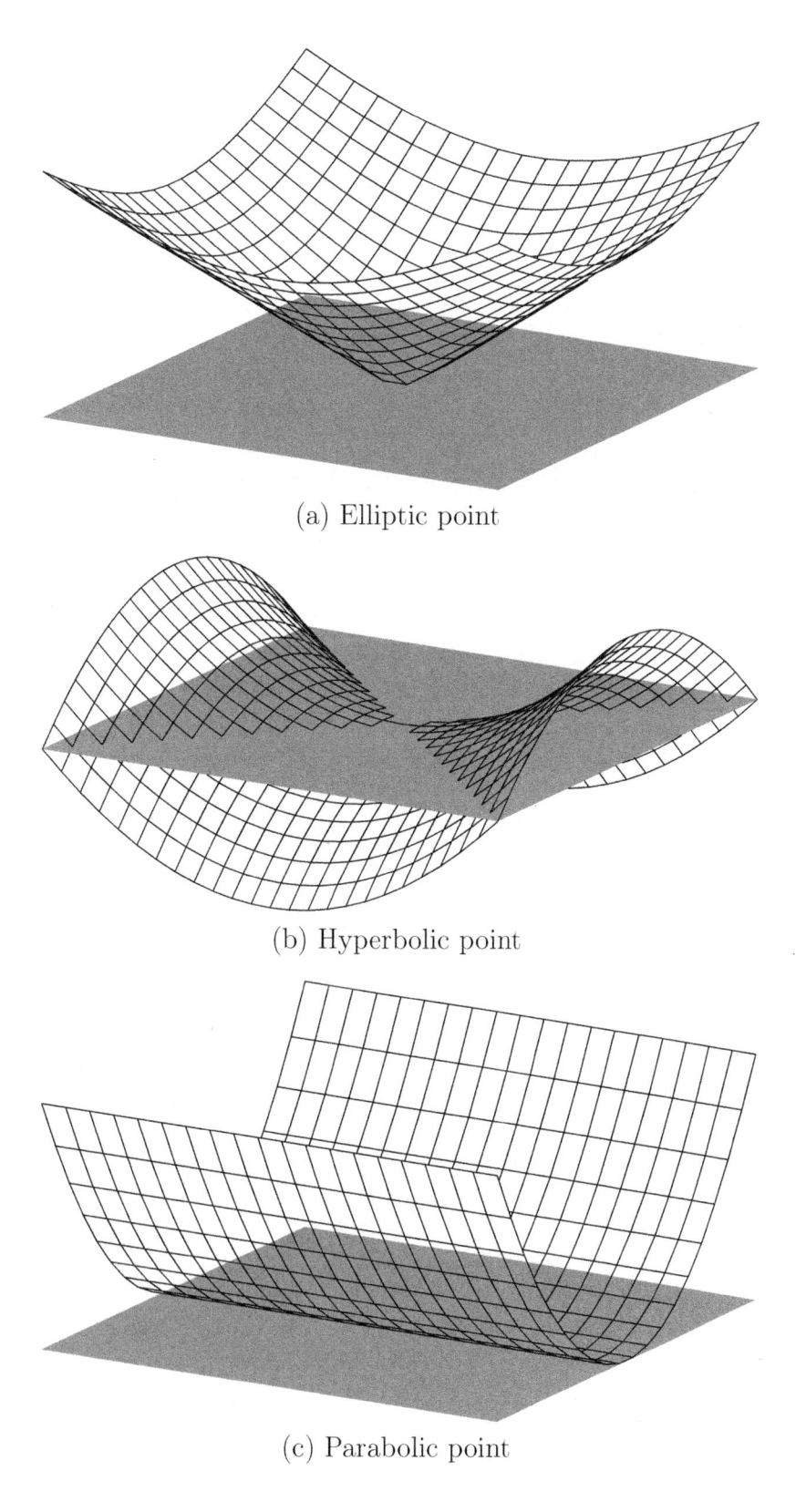

Figure 39: Tangent plane at (a) elliptic point, (b) hyperbolic point and (c) parabolic point.

The surface points can also be classified according to the geometric shape of Dupin indicatrix (refer to § 3.6.1) as follows:

1. If $eg - f^2 = 0$ and $e = f = g = 0$, then the point is flat and the Dupin indicatrix is not defined. Hence, having undefined Dupin indicatrix is a characteristic for planar points. The normal curvature at the point is zero in all directions.
2. If $eg - f^2 = 0$ and $e^2 + f^2 + g^2 > 0$, then either $\kappa_1 = 0$ and $\kappa_2 \neq 0$ or $\kappa_2 = 0$ and $\kappa_1 \neq 0$; hence the point is parabolic and the Dupin indicatrix becomes two parallel lines. The point is characterized by having a vanishing normal curvature along the direction of these lines while it has the same sign in all other directions.
3. If $eg - f^2 > 0$ then $\kappa_1$ and $\kappa_2$ have the same sign; hence the point is elliptic and the Dupin indicatrix is an ellipse or circle. The normal curvature at the point has the same sign in all directions.
4. If $eg - f^2 < 0$ then $\kappa_1$ and $\kappa_2$ have opposite signs; hence the point is hyperbolic and the Dupin indicatrix becomes two conjugate hyperbolas. The normal curvature at the point is positive along the directions corresponding to one of these hyperbolas and negative along the directions corresponding to the other hyperbola, while along the common asymptotes of these hyperbolas the normal curvature is zero.

In brief, because of these correlations between the type of point and the shape of its Dupin indicatrix, the Dupin indicatrix can be used to classify the point as flat, parabolic, elliptic or hyperbolic. It is worth noting that the relation between $eg - f^2$ and $\kappa_1$ and $\kappa_2$ as stated in the above bullet points can be concluded from the fact that (see Eqs. 355 and 356):

$$K = \kappa_1 \kappa_2 = \frac{eg - f^2}{a} \tag{399}$$

since $a$ is positive definite.

We remark that in the immediate neighborhood of a point on a surface, the surface may be approximated by:

1. A plane at a flat point.
2. A parabolic cylinder (Fig. 9) at a parabolic point.
3. An elliptic paraboloid (Fig. 7) at an elliptic point.
4. A hyperbolic paraboloid (Fig. 8) at a hyperbolic point.

Another remark is that in the neighborhood of a parabolic point $P$ on a surface $S$, the tangent plane of $S$ at $P$ meets $S$ in a single line passing through $P$, while in the neighborhood of a hyperbolic point $P$ on a surface $S$, the tangent plane meets $S$ in two lines intersecting at $P$ where these two lines divide $S$ alternatively into regions above the tangent plane and regions below the tangent plane. Finally, the following function:

$$\frac{II_S}{2} = \frac{e\,du\,du + 2f\,du\,dv + g\,dv\,dv}{2} \tag{400}$$

evaluated at a given point $P$ of a class $C^2$ surface may be called the osculating paraboloid of $P$. This osculating paraboloid, represented by half the second fundamental form, is used to determine the shape of the surface at $P$ (also see § 4.2).

## 4.10 Umbilical Point

A point on a surface is called "umbilical" or "umbilic" or "navel" if all the normal sections of the surface at the point have the same normal curvature $\kappa_n$. Hence, at umbilical points we have the following condition:

$$\kappa_1 = \kappa_2 \tag{401}$$

As stated before, for normal sections the normal curvature $\kappa_n$ is equal in magnitude to the curvature $\kappa$. Therefore, the curvature of all the normal sections at umbilical points is also equal. The condition of Eq. 401 implies that an umbilical point cannot be a hyperbolic point because at hyperbolic points we should have $\kappa_1 > 0$ and $\kappa_2 < 0$ (refer to § 4.9). Hence, at umbilical points the Gaussian curvature should satisfy the necessary (but not sufficient) condition: $K \geq 0$.

The following are some examples of umbilical points on common surfaces:
1. All points of planes are umbilical. However, some authors impose the condition $K > 0$ at umbilical points and hence the points of planes are not umbilical according to these authors.
2. All points of spheres are umbilical. Hence, umbilical points may be called spherical points.
3. The vertex of an elliptic paraboloid of revolution is an umbilical point.
4. The two vertices of an ellipsoid of revolution are umbilical points.

If all points of a surface of class $C^3$ are umbilical then the surface must be a sphere. The plane is a special case of sphere as it can be regarded as a sphere with an infinite radius.

A sufficient and necessary condition for a given point $P$ to be umbilical is that the coefficients of the curvature tensor $b_{\alpha\beta}$ at $P$ are proportional to the corresponding coefficients of the metric tensor $a_{\alpha\beta}$ at $P$, that is:

$$b_{\alpha\beta} = c\, a_{\alpha\beta} \qquad (\alpha, \beta = 1, 2) \tag{402}$$

where $c$, which is a proportionality factor, is independent of the direction of the tangent to the normal section at the umbilical point. In fact, this condition is the same as the previously stated condition of Eq. 321, and hence the same justification of Eq. 321 will apply here. We also note that $c = \kappa_n$, as seen there.

As a result of Eq. 402, at umbilical points the determinants of the two tensors, $a$ and $b$, satisfy the following relation:

$$b = c^2 a \tag{403}$$

where $c$ is squared because $a_{\alpha\beta}$ is a tensor represented by a $2 \times 2$ matrix. Now, since the first fundamental form is positive definite, and hence $a > 0$, then if at the umbilical point $c = 0$ then $b = 0$ according to Eq. 403 and the point is a flat umbilic; otherwise $b > 0$ (since $c$ is real) and the point is an elliptic umbilic (see § 4.9).[25] On a plane surface all points are flat umbilic, while on a sphere all points are elliptic umbilic.

As seen earlier, a hyperbolic point cannot be an umbilical point since at the umbilical point we should have identical normal curvatures in all directions, both in sign and in

---

[25] The latter may also be called spherical umbilic.

magnitude, and this cannot happen at a hyperbolic point whose $\kappa_1$ and $\kappa_2$ should be of opposite signs. This is inline with the fact that at an umbilical point $b$ cannot be negative since $c$ is real. In fact, Eq. 403 can be recast into the following form:

$$K = \frac{b}{a} = c^2 = \kappa_n^2 \qquad (404)$$

where the equation: $c = \kappa_n$ and Eq. 356 are used. Hence, all the above-stated facts about the nature of the umbilical point and the impossibility of being hyperbolic, as represented by the condition $K \geq 0$, are justified.

Because at umbilical points $\kappa_1 = \kappa_2$, we have:

$$K = \kappa_1 \kappa_2 = \kappa_1 \kappa_1 = (\kappa_1)^2 = \left(\frac{2\kappa_1}{2}\right)^2 = \left(\frac{\kappa_1 + \kappa_2}{2}\right)^2 = H^2 \qquad (405)$$

where $K$ and $H$ are the Gaussian and mean curvatures at the point (see § 4.5 and 4.6). This can also be obtained from Eq. 348 where the discriminant of this quadratic equation becomes zero (i.e. $4H^2 - 4K = 0$), since at an umbilical point the two roots are equal, and hence $H^2 = K$.

It should be remarked that the provision $H^2 = K$ is a sufficient and necessary condition for a point at which this condition is satisfied to be umbilical. This can be concluded from the stated requirements in the last paragraph. Another remark is that the relation between $K$ and $H$ at umbilical points, as expressed by Eq. 405, may be stated by some authors in the following disguised form:

$$\left(a^{\alpha\beta} b_{\alpha\beta}\right)^2 = \frac{4}{a}\left(b_{11}b_{22} - b_{12}^2\right) \qquad (\alpha, \beta = 1, 2) \qquad (406)$$

where Eqs. 356 and 383 are employed in this form.

## 4.11 Exercises

4.1 Discuss the similarities and differences between the curvature of curves and the curvature of surfaces.
4.2 Define, descriptively and mathematically, the curvature vector $\mathbf{K}$ of surface curves and its relation to the principal normal vector $\mathbf{N}$ of the curve.
4.3 Compare the vectors $\mathbf{n}$ and $\mathbf{N}$ at a point on a surface curve outlining their similarities and differences.
4.4 Discuss the dependency of the curvature vector of a surface curve at a given point of the curve on the following parameters: curve orientation, curve parameterization, surface orientation as indicated by the direction of $\mathbf{n}$, surface parameterization, tangential direction and position of the point on the surface.
4.5 What "inflection point" on a surface curve means?
4.6 What is the radius of curvature at a point of inflection?
4.7 Resolve the curvature vector of a surface curve into its tangential and normal components and name these components. Express these components in terms of the unit vectors $\mathbf{n}$ and $\mathbf{u}$ explaining all the symbols involved in this expression.

## 4.11 Exercises

4.8 Find the curvature vector, $\mathbf{K}$, of a space curve represented by: $\mathbf{r}(t) = (3t^2, t, 2\sin t)$.

4.9 Define, descriptively and quantitatively, the normal and geodesic curvatures $\kappa_n$ and $\kappa_g$.

4.10 Which of $\kappa_n$ and $\kappa_g$ is an intrinsic property and which is an extrinsic property? Explain why.

4.11 Compare the following four moving frames: $(\mathbf{T}, \mathbf{N}, \mathbf{B})$, $(\mathbf{E}_1, \mathbf{E}_2, \mathbf{n})$, $(\mathbf{n}, \mathbf{T}, \mathbf{u})$ and $(\mathbf{d}_1, \mathbf{d}_2, \mathbf{n})$ outlining their similarities and dissimilarities.

4.12 Which of the frames in the previous question employ both surface and curve vectors? Which of these frames are orthonormal by definition and which are not?

4.13 How the curvature $\kappa$ of a surface curve at a given point is related to its normal and geodesic curvatures $\kappa_n$ and $\kappa_g$ at that point? Can you make sense of this considering the normal and tangential components of the curvature vector $\mathbf{K}$?

4.14 Prove that the geodesic curvature of a naturally parameterized curve is given by Eq. 325.

4.15 Show that in any two orthogonal directions at a given point $P$ on a sufficiently smooth surface, the sum of the normal curvatures corresponding to these directions at $P$ is constant.

4.16 Give a brief statement of the theorem of Meusnier outlining its significance. State this theorem in a second alternative form.

4.17 Show that the osculating circles of all curves on a surface that pass through a given point and in a specific direction are on a sphere.

4.18 Define, descriptively and mathematically, the normal component $\mathbf{K}_n$ of the curvature vector $\mathbf{K}$ of a surface curve outlining its relation to the curvature vector and the normal vector to the surface, $\mathbf{n}$.

4.19 Define, descriptively and mathematically, the geodesic component $\mathbf{K}_g$ of the curvature vector $\mathbf{K}$ of a surface curve outlining its relation to the curvature vector and the surface basis vectors $\mathbf{E}_1$ and $\mathbf{E}_2$.

4.20 Derive the formula for the normal curvature $\kappa_n$ as a ratio of the second fundamental form to the first fundamental form.

4.21 Why the sign of the normal curvature $\kappa_n$ is determined only by the sign of the second fundamental form?

4.22 Show that at any point $P$ of a smooth surface $S$ there exists a paraboloid tangent to $S$ at $P$ such that the normal curvature of the paraboloid in any direction is equal to the normal curvature of $S$ at $P$ in that direction.

4.23 Discuss, in detail, the following statement: "The normal curvature at a given point on a surface and in a given tangential direction to the surface is a property of the surface". Can you link this to the Meusnier theorem?

4.24 For what type of surface curve the following relation is true: $|\kappa_n| = \kappa$? Explain why this is so.

4.25 What is the significance of having a paraboloid at the points of a smooth surface whose normal curvature in a given direction is equal to the normal curvature of the surface in that direction?

4.26 What is the sign of $b$ (i.e. being greater than, less than or equal to zero) at flat,

## 4.11 Exercises

elliptic, parabolic and hyperbolic points on a surface, where $b$ is the determinant of the surface covariant curvature tensor?

4.27 Classify the local shape of a surface at a given point $P$ according to the values of $K$ and $H$ at $P$.

4.28 Using one of the mathematical definitions of the geodesic curvature $\kappa_g$, explain why $\kappa_g$ should be classified as an intrinsic or extrinsic property.

4.29 At what type of surface points the following relation is true: $\frac{e}{E} = \frac{f}{F} = \frac{g}{G} = c$ where $c$ is constant for all directions? What $c$ stands for?

4.30 Express the equalities in the previous question in terms of the coefficients of the covariant metric and covariant curvature tensors, $a_{\alpha\beta}$ and $b_{\alpha\beta}$, of the surface.

4.31 Outline two direct consequences of Meusnier theorem.

4.32 Write a mathematical relation linking the geodesic component $\mathbf{K}_g$ of the curvature vector to the surface basis vectors $\mathbf{E}_1$ and $\mathbf{E}_2$.

4.33 What is the relation between $\mathbf{K}_g$ and the tangent space $T_P S$ of the surface at a given point?

4.34 Give the formulae of the geodesic curvature $\kappa_g$ of the coordinate curves. Simplify these formulae in the case of having orthogonal coordinate curves.

4.35 State a mathematical relation between the curvature $\kappa$ and the geodesic curvature $\kappa_g$ of a surface curve at a given point on the curve explaining all the symbols involved.

4.36 Give a formula for the geodesic curvature $\kappa_g$ in which extrinsic entities are involved. Does this mean that $\kappa_g$ is an extrinsic property of the surface?

4.37 Explain, in detail, all the symbols used in the following formula:

$$\kappa_g = \frac{d\theta}{ds} + \kappa_{gu} \cos\theta + \kappa_{gv} \sin\theta$$

What is the name of this formula?

4.38 Prove the relation given in the last question.

4.39 Give a mathematical formula in which $\kappa_n$ is expressed in terms of the coefficients of the first and second fundamental forms $E, F, G, e, f, g$.

4.40 What the two "principal curvatures" of a surface at a given point mean?

4.41 Find analytical expressions for the principal curvatures on a surface represented by the equation: $\xi_2 \cos\xi_3 - \xi_1 \sin\xi_3 = 0$ where $\xi_1, \xi_2, \xi_3$ are real variables.

4.42 The principal curvatures of a surface at a given point correspond to the two directions represented by $\lambda_1$ and $\lambda_2$ which are the roots of the following quadratic equation:

$$(gF - fG)\lambda^2 + (gE - eG)\lambda + (fE - eF) = 0$$

From the rules of polynomial equations, find the sum and product of these roots.

4.43 Define the "principal directions" descriptively and mathematically.

4.44 Show that $\kappa$ is a principal curvature with a principal direction $\frac{dv}{du}$ iff the following conditions are satisfied:

$$(e - \kappa E)du + (f - \kappa F)dv = 0$$
$$(f - \kappa F)du + (g - \kappa G)dv = 0$$

## 4.11 Exercises

4.45 Find the principal curvatures and the principal directions on a surface represented parametrically by: $\mathbf{r}(u,v) = (u, v, 2u^2 + 5v^2)$ at the point with $(u,v) = (2.3, 1.6)$.

4.46 Prove Euler theorem (see Eq. 341 and the surrounding text).

4.47 What is Darboux frame? Are the vectors of this frame orthonormal? Is this frame defined at umbilical points on the surface? Fully justify your answer related to the last two parts of the question.

4.48 Write the formulae for the positions of the centers of curvature of the normal sections corresponding to the two principal curvatures at a given point on a surface.

4.49 Correlate, mathematically with full explanation of all the symbols involved, the normal curvature $\kappa_n$ at a given point and in a given direction on a smooth surface to the two principal curvatures at that point.

4.50 Define, mathematically in terms of the principal curvatures, the following terms: principal radii, mean curvature and Gaussian curvature.

4.51 Distinguish between the "total curvature" of a curve and the "total curvature" of a surface. For surface, what are the two meanings of this term?

4.52 Find the Gaussian and mean curvatures of a surface given by: $\mathbf{r}(u,v) = (3u - v, u + 2v, 1.5uv)$ at the point with $(u,v) = (3,1)$.

4.53 State the limiting conditions on the principal curvatures, and hence deduce the conditions on the mean and Gaussian curvatures, on the surface of sphere and on the surface of hyperboloid of one sheet.

4.54 Prove that there is no compact surface of class $C^2$ with non-positive Gaussian curvature over the whole surface.

4.55 Analyze the following equation outlining its significance:

$$S(x,y) \simeq S(0,0) + \frac{\kappa_1 x^2}{2} + \frac{\kappa_2 y^2}{2}$$

4.56 State the necessary and sufficient condition for a real number to be a principal curvature of a surface at a given point.

4.57 Investigate the number of roots of the following quadratic equation and the impact of this on the number of principal curvatures of the surface at the point where this equation applies:

$$\left(EG - F^2\right)\kappa^2 - \left(gE - 2fF + eG\right)\kappa + \left(eg - f^2\right) = 0$$

4.58 From the equation in the previous question, obtain an analytical expression for the principal curvatures of the surface at the point where this equation applies.

4.59 From the equation in the last two questions, obtain the equation: $\kappa^2 - 2H\kappa + K = 0$ and hence verify that the principal curvatures are given by: $\kappa_{1,2} = H \pm \sqrt{H^2 - K}$.

4.60 Write down the equations of the principal curvatures when the $u^1$ and $u^2$ coordinate curves are aligned along the principal directions.

4.61 Obtain Eq. 367 for the Gaussian curvature of a surface with orthogonal coordinate curves by using Eq. 366.

4.62 Show that the spheres are the only connected, compact and sufficiently smooth surfaces with constant Gaussian curvature.

## 4.11 Exercises

4.63 State the curvature formula of Rodrigues defining all the symbols involved and discussing its significance.

4.64 Test the validity of the Rodrigues formula for the principal directions at the point with $(u, v) = (1.4, 3.9)$ on a surface parameterized by: $\mathbf{r}(u, v) = (u, v, u^2 + 3v^2)$.

4.65 Use the Rodrigues curvature formula to prove that spheres are the only connected closed surfaces of class $C^3$ whose all points are spherical umbilical.

4.66 Give a mathematical expression for the Gaussian curvature in terms of the coefficients of the surface metric and curvature tensors.

4.67 What is the significance of having an intrinsic surface curvature, represented usually by the Gaussian curvature, as a way for a 2D inhabitant to have some perception of the nature of the surface and its shape as seen from the ambient space by a 3D inhabitant?

4.68 Discuss the following statement: "The Gaussian curvature along any parallel line of a surface of revolution is constant".

4.69 Starting from the following relation: $K = \frac{b}{a}$, derive the relation: $K = \det(b_\beta^\alpha)$.

4.70 Give a mathematical relation correlating the Gaussian curvature to the following coefficients of the 2D Riemann-Christoffel curvature tensor: $R_{1212}$, $R_{1221}$, $R_{1121}$ and $R_{2112}$.

4.71 What is the Gaussian curvature of a Monge patch of the form $\mathbf{r}(u, v) = (u, v, f(u, v))$?

4.72 Why the Gaussian curvature is independent of the orientation of the surface (where orientation is based on the choice of the direction of the unit normal vector to the surface)?

4.73 Which of the following geometric shapes have identical Gaussian curvatures at their corresponding points and why: plane, sphere, cylinder, catenoid, ellipsoid, hyperbolic paraboloid, helicoid, and cone? Compare, in your answer, each pair of these shapes.

4.74 Write down an expression for the Gaussian curvature of a surface of revolution generated by revolving a sufficiently differentiable plane curve of the form $x = f(y)$ around the $y$-axis.

4.75 Explain all the symbols of the following equation with discussion of its significance in relation to the intrinsic and extrinsic geometries of the surface:

$$\partial_u \mathbf{n} \times \partial_v \mathbf{n} = K \left( \mathbf{E}_1 \times \mathbf{E}_2 \right)$$

4.76 State the mathematical expression that correlates the Gaussian curvature to the Ricci curvature scalar of a surface.

4.77 Using Eq. 367 and the parametric equations of Beltrami pseudo-sphere (Eqs. 43-45), show that the pseudo-sphere has a negative constant Gaussian curvature and find this curvature.

4.78 Classify surfaces with regard to their Gaussian curvature as having constant or variable curvature giving two examples for each.

4.79 What is the impact of scaling a surface up or down by a constant positive factor on its Gaussian curvature?

4.80 Discuss the effect of an isometric mapping of a surface on its Gaussian curvature.

4.81 State the Hilbert lemma giving examples for its applications from common types of

## 4.11 Exercises

surface.

4.82 Give the conditions for the validity of the following equation:

$$K = -\frac{1}{2\sqrt{EG}} \left[ \partial_u \left( \frac{G_u}{\sqrt{EG}} \right) + \partial_v \left( \frac{E_v}{\sqrt{EG}} \right) \right]$$

Also, give its simplified form in the case of representing the surface by geodesic coordinates stating the other conditions required for this simplification.

4.83 Express the mean curvature as a function of the Gaussian curvature taking care of the signs.

4.84 Show that spheres are the only connected compact surfaces with constant mean curvature and positive Gaussian curvature.

4.85 Explain how the position of the surface in a deleted neighborhood of a given point $P$ relative to the tangent plane of the surface at $P$ is used to classify the nature of the Gaussian curvature at $P$. From this perspective, discuss the sign of the Gaussian curvature on the points of the following surfaces: hyperbolic paraboloid, sphere, torus and cylinder.

4.86 The Gaussian curvature of a developable surface is identically zero. Why?

4.87 What is the Gaussian curvature of a surface parameterized by: $x = (5 + \cos\phi)\cos\theta$, $y = (5 + \cos\phi)\sin\theta$ and $z = \sin\phi$?

4.88 Provide a mathematical definition for the total curvature of a surface explaining all the symbols used in the definition.

4.89 Define all the symbols used in the following equation:

$$\underline{\epsilon}^{\alpha\beta}\underline{\epsilon}^{\gamma\delta}R_{\alpha\beta\gamma\delta} = K\underline{\epsilon}^{\alpha\beta}\underline{\epsilon}^{\gamma\delta}\underline{\epsilon}_{\alpha\beta}\underline{\epsilon}_{\gamma\delta}$$

4.90 Explain in detail how the following equation implies that the Gaussian curvature is a rank-0 tensor: $K = \frac{1}{4}\underline{\epsilon}^{\alpha\beta}\underline{\epsilon}^{\gamma\delta}R_{\alpha\beta\gamma\delta}$.

4.91 Write the Gaussian curvature in terms of the surface curvature tensor using the most simple form.

4.92 Algebraically manipulate the relation $K = \frac{b}{a}$ to obtain the following relation:

$$K = \frac{(\partial_u \mathbf{E}_1 \cdot \mathbf{E}_1 \times \mathbf{E}_2)(\partial_v \mathbf{E}_2 \cdot \mathbf{E}_1 \times \mathbf{E}_2) - (\partial_v \mathbf{E}_1 \cdot \mathbf{E}_1 \times \mathbf{E}_2)^2}{(EG - F^2)^2}$$

4.93 Express the mean curvature $H$ in terms of the coefficients of the first and second fundamental forms.

4.94 What is the relation between the mean curvature $H$ and the mixed type surface curvature tensor $b_\alpha^\beta$?

4.95 Compare the sign of the mean curvature to the sign of the Gaussian curvature with regard to their dependency on the direction of the unit normal vector to the surface.

4.96 Give two examples of common types of surface over which the mean curvature is constant. Also, give an example of a surface with variable mean curvature.

4.97 What is the mean curvature of a Monge patch of the form $\mathbf{r}(u,v) = (u, v, f(u,v))$?

## 4.11 Exercises

4.98 What is the essence of Gauss *Theorema Egregium*? Give an example of an equation or a theorem that demonstrates this *theorem*.

4.99 Derive Eq. 385 using Eq. 380.

4.100 Write down the mathematical equation representing the local form of the Gauss-Bonnet theorem explaining all the symbols involved.

4.101 Give an example for the application of the local Gauss-Bonnet theorem using a planar geometric shape and another example using a non-planar shape.

4.102 Explain why two geodesic curves on a patch of a surface with negative Gaussian curvature cannot intersect at two points.

4.103 Apply the Gauss-Bonnet theorem on the spherical triangle of Fig. 37 giving detailed explanations for each step.

4.104 Use a circular flat disc to demonstrate the application of the local form of the Gauss-Bonnet theorem giving detailed explanations for each step.

4.105 What is the global form of the Gauss-Bonnet theorem and what is its significance geometrically and topologically?

4.106 Find the total curvature of the surfaces depicted in Fig. 17.

4.107 Show, mathematically, that the area of a geodesic polygon on a surface with constant non-vanishing Gaussian curvature is determined by the sum of the internal angles of the polygon.

4.108 Verify that the total curvatures of ellipsoid and torus are respectively $4\pi$ and $0$ by performing detailed surface integral calculations.

4.109 Outline the usefulness of the global form of the Gauss-Bonnet theorem in obtaining the total curvature of a surface with known topological properties without performing detailed calculations.

4.110 Using the Gauss-Bonnet theorem, prove that the Gaussian curvature is identically zero on a surface $S$ if at any point $P$ on $S$ there are two families of geodesic curves in the neighborhood of $P$ intersecting at a constant angle.

4.111 Write down the mathematical relation that links the Euler characteristic of a surface to its topological genus.

4.112 Use the principal curvatures and the mean and Gaussian curvatures to classify the points with regard to the local shape of the surface as flat, elliptic, parabolic and hyperbolic giving examples of common geometric shapes for each case.

4.113 Repeat the classification of the previous question using this time the coefficients of the second fundamental form of the surface.

4.114 Prove that on a circular cylinder all points are parabolic.

4.115 Show that in the neighborhood of an elliptic point on a surface, the surface lies on one side of its tangent plane at that point.

4.116 A surface is represented parametrically by: $\mathbf{r}(u,v) = (u, v, u^2 + v^3)$. Determine the conditions that identify the parabolic, hyperbolic and elliptic points on the surface.

4.117 Give an example of a surface having elliptic, parabolic and hyperbolic points at different locations.

4.118 Why the point type (i.e. being flat, elliptic, hyperbolic or parabolic) on a surface is an invariant property with respect to changes in the surface representation and

## 4.11 Exercises

parameterization?

4.119 Why the point type is invariant with respect to a change of the surface orientation by reversing the direction of the normal vector to the surface?

4.120 Make a simple sketch outlining the position of a surface relative to the tangent plane at elliptic, parabolic and hyperbolic tangency points.

4.121 Demonstrate that the surface represented parametrically by: $\mathbf{r}(u,v) = (u, v, u^2 + v^3)$ lies on both sides of its tangent plane at the point $(u, v) = (0, 0)$.

4.122 For a surface represented parametrically by: $\mathbf{r} = (u, v, v^4)$, find the equation of a curve on the surface whose points have a common tangent plane.

4.123 Describe how Dupin indicatrix can be used to classify the points of a surface with regard to the local shape (i.e. flat, elliptic, parabolic and hyperbolic).

4.124 What are the prototypical geometric shapes that provide the best approximation for the local shape of a sufficiently smooth surface at its: flat, elliptic, hyperbolic and parabolic points?

4.125 What "umbilical point" means? What are the other terms used to label such a point?

4.126 What are the characteristic features of umbilical points?

4.127 Give five examples of umbilical points on common geometric surfaces such as spheres and paraboloids.

4.128 State the mathematical relation between the coefficients of the metric and curvature tensors at umbilical points.

4.129 Demonstrate that at an umbilical point of a surface we have: $K = H^2$ where $K$ and $H$ are the Gaussian and mean curvatures at the point.

4.130 Show that the relation: $K = H^2$ can also be written as:

$$\left(a^{\alpha\beta}b_{\alpha\beta}\right)^2 = \frac{4}{a}\left(b_{11}b_{22} - b_{12}^2\right)$$

4.131 Explain why at umbilical points we have $b = c^2 a$ where $a$ and $b$ are the determinants of the covariant metric and covariant curvature tensors and $c$ is a proportionality factor.

4.132 Give two examples of surfaces whose all points are umbilical, and two other examples of surfaces with no umbilical point at all. Also, give an example of a surface with only one umbilical point, and another example of a surface with only two umbilical points.

# Chapter 5
# Special Curves

There are many classifications to space curves depending on their properties and their relations with each other. In the following sections of this chapter, we briefly investigate a few of these categories.

## 5.1 Straight Line

A necessary and sufficient condition for a curve of class $C^2$ to be a straight line is that its curvature is zero at every point on the curve. Hence, another criterion for a curve to be a straight line is that all the tangents of the curve are parallel, where "parallel" here is used in its absolute Euclidean sense (see § 2.7). Another criterion for a curve $C(t) : I \to \mathbb{R}^3$ where $t \in I \subseteq \mathbb{R}$ to be a straight line is that for all points $t$ in the domain of the curve, $\dot{\mathbf{r}}$ and $\ddot{\mathbf{r}}$ are linearly dependent where $\mathbf{r}(t)$ is the spatial representation of the curve and the overdots represent derivative with respect to the general parameter $t$ of the curve. A straight line lying on a surface has the same tangent plane at each of its points, and hence the line is contained in this unique tangent plane. Any straight line on any surface is a geodesic curve (see § 5.7) and an asymptotic line (see § 5.9).

## 5.2 Plane Curve

A curve is described as a plane curve if the whole curve can be contained in a plane with no distortion (Fig. 40). A necessary and sufficient condition for a curve parameterized by a general parameter $t$ to be a plane curve is that the relation $\dot{\mathbf{r}} \cdot (\ddot{\mathbf{r}} \times \dddot{\mathbf{r}}) = 0$ holds identically where $\mathbf{r}(t)$ is the spatial representation of the curve and the overdots represent differentiation with respect to $t$. Plane curves are characterized by having identically vanishing torsion. In fact, having identically vanishing torsion is a necessary and sufficient condition for a regular curve of class $C^2$ to be a plane curve. For plane curves, the osculating plane at each regular point on the curve contains the entire curve. Therefore, the plane curve may be characterized by having a common intersection point for all of its osculating planes. It also implies that the curve has the same osculating plane at all of its points. Two curves are plane curves if they have the same binormal lines at each pair of their corresponding points. The locus of the centers of curvature of a curve $C$ is an evolute (see § 5.3) of $C$ *iff* $C$ is a plane curve. On a smooth surface, a geodesic curve (see § 5.7) which is also a line of curvature (see § 5.8) is a plane curve. A plane curve has always a Bertrand curve associate (see § 5.4).

## 5.3 Involute and Evolute

Figure 40: Plane curve.

### 5.3 Involute and Evolute

If $C_e$ is a space curve with a tangent surface $S_T$ (see § 6.6) and $C_i$ is a curve embedded in $S_T$ and it is orthogonal to all the tangent lines of $C_e$ at their intersection points, then $C_i$ is called an involute of $C_e$ while $C_e$ is called an evolute of $C_i$ (see Fig. 41). Hence, the involute is an orthogonal trajectory of the generators of the tangent surface of its evolute. Accordingly, the equation of an involute $C_i$ to a curve $C_e$ is given by:

$$\mathbf{r}_i = \mathbf{r}_e + (c - s)\,\mathbf{T}_e \qquad (407)$$

where $\mathbf{r}_i$ is an arbitrary point on the involute, $\mathbf{r}_e$ is the point on the curve $C_e$ corresponding to $\mathbf{r}_i$, $c$ is a given constant, $s$ is a natural parameter of $C_e$ and $\mathbf{T}_e$ is the unit vector tangent to $C_e$ at $\mathbf{r}_e$.

A visual demonstration of how to generate an involute $C_i$ of a curve $C_e$, when $(c-s)$ in Eq. 407 is positive, may be given by detaching a taut string attached to $C_e$ where the string is kept in the tangent direction as it is detached. A fixed point $P$ on the string, where the distance between $P$ and the point of contact of the string with $C_e$ represents a natural parameter of $C_e$, then traces an involute of $C_e$.

A curve has infinitely many involutes corresponding to different values of $c$ in Eq. 407. Therefore, the involutes may be described as parallel curves on the tangent surface. Similarly, an involute has an infinite number of evolutes corresponding to different values of $c$. For any tangent of a given curve, the length of the line segment confined between two given involutes is constant which is the difference between the two $c$'s in Eq. 407 of the two involutes.

If $C_e$ is an evolute of $C_i$, then for a given point $P_e$ on $C_e$ and the corresponding point $P_i$ on $C_i$ the principal normal line of $C_e$ at $P_e$ is parallel to the tangent line of $C_i$ at $P_i$. A curve $C_i$ is a plane curve *iff* the locus of the centers of curvature of $C_i$ is an evolute of $C_i$. The involutes of a circle are congruent. The evolutes of plane curves are helices.

### 5.4 Bertrand Curve

Bertrand curves are two associated space curves with common principal normal lines at their corresponding points. Associated Bertrand curves are characterized by the following properties:

## 5.4 Bertrand Curve

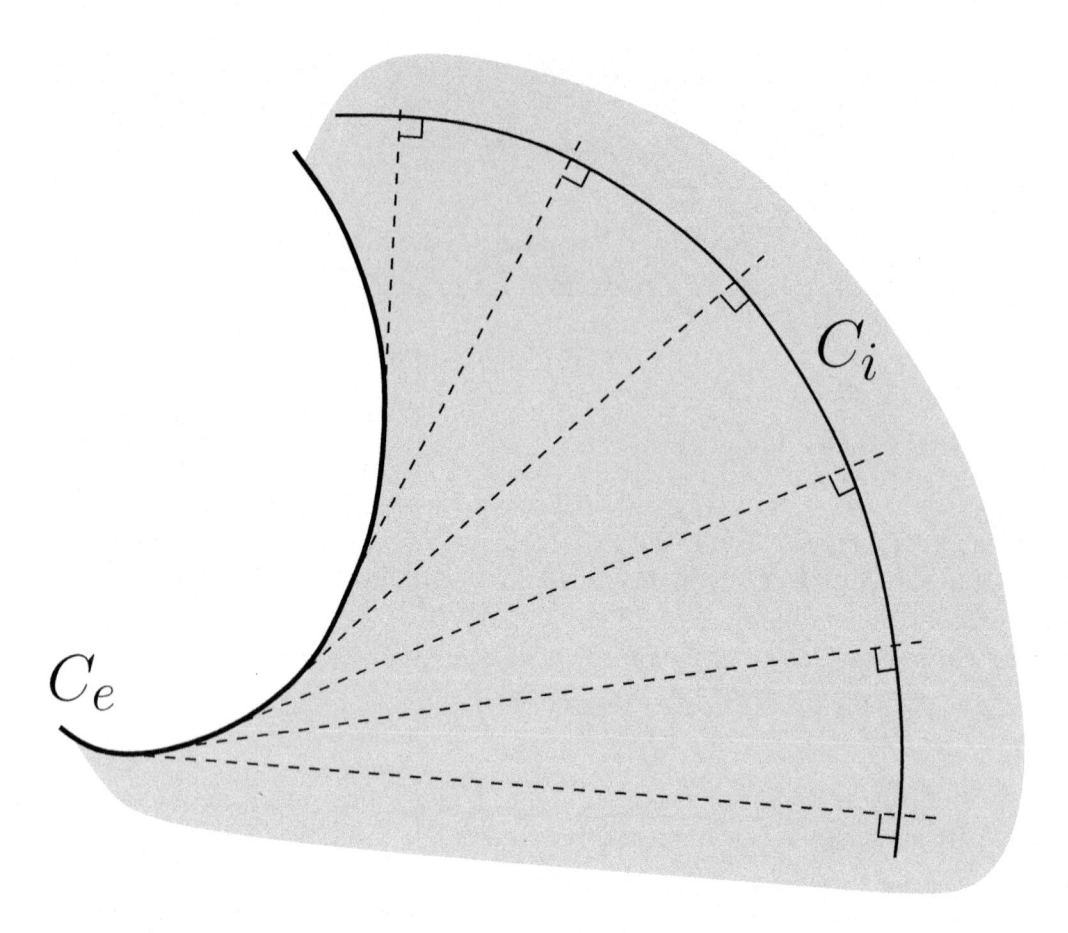

Figure 41: Evolute $C_e$, involute $C_i$, tangent lines (dashed) and tangent surface (shaded).

1. The product of the torsions of their corresponding points is constant, that is:

$$\tau_1(s_o)\tau_2(s_o) = \text{constant} \tag{408}$$

   where $\tau_1$ and $\tau_2$ are the torsions of the two curves and $s_o$ is a given value of their common parameter.
2. The distance between their corresponding points is constant.
3. The angle between their corresponding tangent lines is constant.

For a plane curve $C_1$, there is always a curve $C_2$ such that $C_1$ and $C_2$ are associated Bertrand curves. If $C_1$ is a curve with non-vanishing torsion such that $C_1$ has more than one Bertrand curve associate, then $C_1$ is a circular helix. The reverse is also true. If $C_1$ is a curve with non-vanishing torsion then a necessary and sufficient condition for $C_1$ to be a Bertrand curve (i.e. it possesses an associate curve $C_2$ such that $C_1$ and $C_2$ are Bertrand curves) is that there are two constants $c_1$ and $c_2$ such that:

$$\kappa = c_1\tau + c_2 \tag{409}$$

where $\kappa$ and $\tau$ are the curvature and torsion of the curve $C_1$. If $C_1$ and $C_2$ are two involutes of a plane curve $C$, then $C_1$ and $C_2$ are Bertrand curves.

## 5.5 Spherical Indicatrix

A spherical indicatrix of a continuously-varying unit vector is a continuous curve $\bar{C}$ on the origin-based unit sphere generated by mapping the unit vector (e.g. **T** or **N** or **B**) of a particular space curve $C$ on an equal unit vector represented by a point on the origin-based unit sphere. Hence, we have $\bar{C}_\mathbf{T}$, $\bar{C}_\mathbf{N}$ and $\bar{C}_\mathbf{B}$ as the spherical indicatrices of $C$ corresponding respectively to the tangent, principal normal and binormal vectors of $C$.[26] Figure 42 is a simple demonstration of the spherical indicatrix $\bar{C}_\mathbf{T}$ of a space curve $C$.

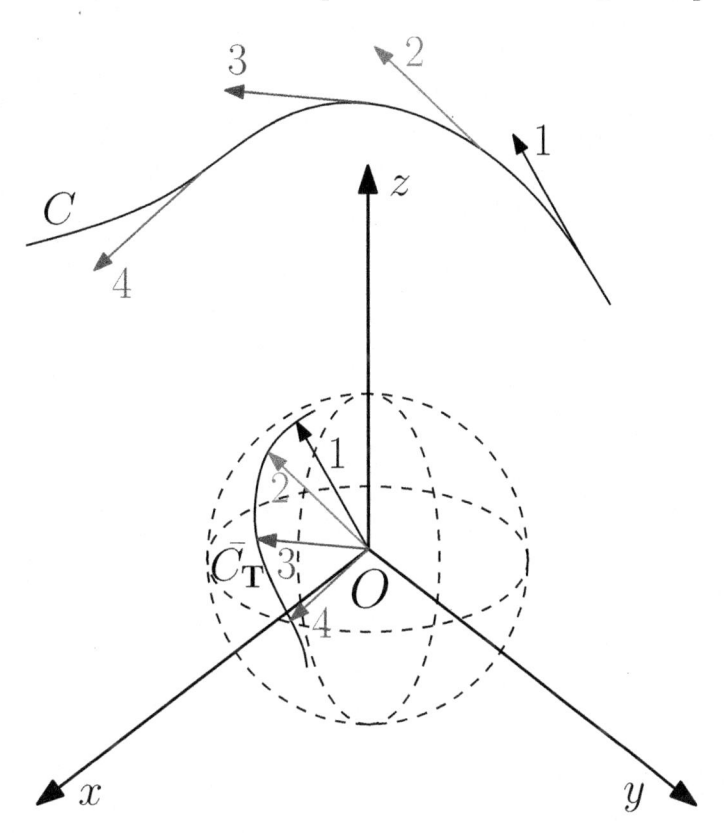

Figure 42: The spherical tangent indicatrix $\bar{C}_\mathbf{T}$ of a space curve $C$ where the numbers indicate the correspondence between the unit tangent vectors of $C$ and their map on $\bar{C}_\mathbf{T}$.

If $C(s)$ is a naturally parameterized curve then $s$ will not necessarily be a natural parameter for the tangent indicatrix $\bar{C}_\mathbf{T}$. A necessary and sufficient condition for $s$ to be a natural parameter for $\bar{C}_\mathbf{T}$ is that $\kappa(s) = 1$ identically where $\kappa$ is the curvature of $C$. The tangent to the curve $\bar{C}_\mathbf{T}$ of a curve $C$ is parallel to the normal vector **N** of $C$ at the corresponding points of the two curves. The tangent to the curve $\bar{C}_\mathbf{T}$ of a curve $C$ is also parallel to the tangent to the curve $\bar{C}_\mathbf{B}$ of $C$ at the corresponding points of the two curves. The necessary and sufficient condition for the curve $\bar{C}_\mathbf{T}$ of a curve $C$ to be a circle is that $C$ is a helix.

---

[26] As seen, the spherical indicatrix may be ascribed to the vector or to the curve; the meaning should be obvious.

The curvature of the curve $\bar{C}_\mathbf{T}$ of a curve $C$ is related to the curvature and torsion of $C$ by:

$$\kappa_\mathbf{T}^2 = \frac{\kappa^2 + \tau^2}{\kappa^2} \tag{410}$$

where $\kappa_\mathbf{T}$ is the curvature of $\bar{C}_\mathbf{T}$ while $\kappa$ and $\tau$ are the curvature and torsion of $C$ respectively. The torsion of the curve $\bar{C}_\mathbf{T}$ of a naturally parameterized curve $C$ is given by:

$$\tau_\mathbf{T} = \frac{\kappa'\tau - \kappa\tau'}{\kappa\left(\kappa^2 + \tau^2\right)} \tag{411}$$

where $\tau_\mathbf{T}$ is the torsion of $\bar{C}_\mathbf{T}$, $\kappa$ and $\tau$ are the curvature and torsion of $C$ respectively, and the prime stands for derivative with respect to the natural parameter $s$ of $C$.

The curvature of the curve $\bar{C}_\mathbf{B}$ of a curve $C$ is given by:

$$\kappa_\mathbf{B} = \frac{\kappa^2 + \tau^2}{\kappa^2} \tag{412}$$

where $\kappa_\mathbf{B}$ is the curvature of $\bar{C}_\mathbf{B}$ while the other symbols are as explained before. The torsion of the curve $\bar{C}_\mathbf{B}$ of a naturally parameterized curve $C$ is given by:

$$\tau_\mathbf{B} = \frac{\kappa'\tau - \kappa\tau'}{\tau\left(\kappa^2 + \tau^2\right)} \tag{413}$$

where $\tau_\mathbf{B}$ is the torsion of $\bar{C}_\mathbf{B}$.

## 5.6 Spherical Curve

A spherical curve is a curve that lies completely on the surface of a sphere. Spherical indicatrices are common examples of spherical curves (see § 5.5). Circles are the only spherical curves with constant curvature. At all points of a spherical curve, the normal plane of the curve passes through the center of the embedding sphere. Conversely, if all the normal planes of a curve meet in a common point, then the curve is spherical with the common point being the center of the sphere that envelops the curve.

The sufficient and necessary condition that should be satisfied by a spherical curve is given by:

$$\frac{R_\kappa}{R_\tau} + \frac{d}{ds}\left(R_\tau \frac{dR_\kappa}{ds}\right) = 0 \tag{414}$$

where $R_\kappa$ and $R_\tau$ are the radii of curvature and torsion and $s$ is a natural parameter of the curve. The center of curvature of a twisted spherical curve $C$ at a given point $P$ on $C$ is the projection of the center of the enveloping sphere on the osculating plane of $C$ at $P$.

## 5.7 Geodesic Curve

The characteristic feature of a geodesic curve is that it has vanishing geodesic curvature $\kappa_g$ at every point on the curve. This is a necessary and sufficient condition for a surface

## 5.7 Geodesic Curve

curve to be geodesic. In more technical terms, let $S : \Omega \to \mathbb{R}^3$ be a surface defined on a set $\Omega \subseteq \mathbb{R}^2$ and let $C(t) : I \to \mathbb{R}^3$, where $I \subseteq \mathbb{R}$, be a regular curve on $S$, then $C$ is a geodesic curve *iff* $\kappa_g(t) = 0$ on all points $t \in I$ in its domain. The path of the shortest distance connecting two points in a Riemannian space is a geodesic. The length of arc, as given by Eq. 204, is used in the definition of geodesic in this sense.

A physical interpretation may be given to the geodesic curve that a free particle restricted to move on the surface will follow a geodesic path. Another physical interpretation is that a geodesic path minimizes the total kinetic energy spent by a massive object in moving between two points when the path is traversed with constant speed. These two physical interpretations may rest on the same physical principle.

The geodesic is a straight line in a Euclidean space, but it is a generalized curved path in a general Riemannian space. If a geodesic surface curve is not a straight line then its principal normal vector **N** is collinear with the normal vector **n** to the surface at each point on the curve with non-vanishing curvature; the opposite is also true. In fact, a curve on a surface is geodesic *iff* it is either a straight line or its principal normal vector is collinear with **n** over the whole curve. As stated before, collinearity of **N** and **n** is equivalent to the condition that **n** lies in the osculating plane of the curve at the given point.

Another sufficient and necessary condition for a curve to be a geodesic curve is that the first variation (see § 1.4.2) of its length is zero. In fact, this may be taken as the basis for the definition of geodesic as the curve connecting two fixed points, $P_1$ and $P_2$, whose length possesses a stationary value with regard to small variations in its neighborhood, that is:

$$\delta \int_{P_1}^{P_2} ds = 0 \tag{415}$$

It can be shown that a geodesic curve satisfies the Euler-Lagrange variational principle (see § 1.4.2) which is a necessary and sufficient condition for extremizing the arc length. Figure 43 is an illustration of how the length of the geodesic curve between two given points is subject to the variational principle.

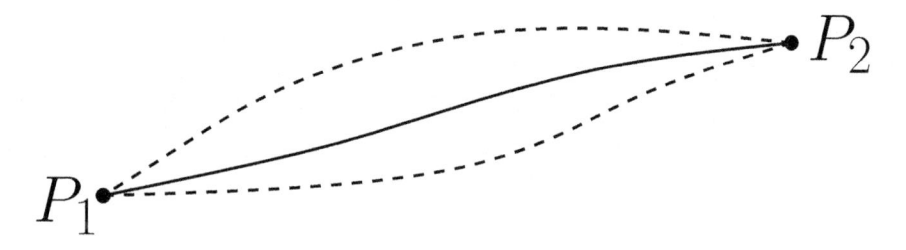

Figure 43: The length of a geodesic curve (solid) connecting two points, $P_1$ and $P_2$, as an extremum with respect to the length of other curves (dashed) connecting these points that result from small perturbations in its neighborhood.

Examples of geodesic curves on simple surfaces are the arcs of great circles on spheres. In fact, being an arc of a great circle is a sufficient and necessary condition for being a geodesic curve on a sphere. Other examples of geodesic curves are the arcs of helices, the

## 5.7 Geodesic Curve

generating straight lines and the circles on cylinders (Fig. 44). The generating straight lines and the circles on cylinders may be considered as degenerate helices. The meridians of a surface of revolution are also geodesics. The arcs of parallel circles on a surface of revolution corresponding to stationary points on the generating curve of the surface are also geodesic curves. All straight lines on any surface are geodesic curves. For plane surfaces in particular, being a straight line on a plane is a sufficient and necessary condition for being a geodesic. The lines of curvature (see § 5.8) are also geodesic curves.

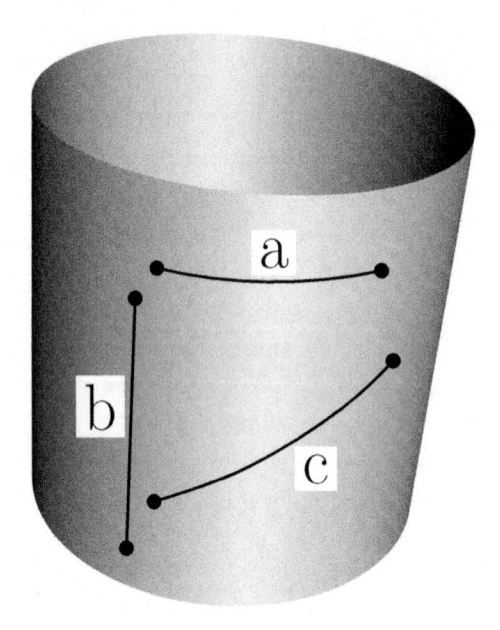

Figure 44: The three types of geodesic curves on cylinders (a) circular arcs (b) generating lines and (c) helical arcs.

Intrinsically, the geodesic curves are straight lines in the sense that a 2D inhabitant will see them straight since he cannot detect their curvature. This is due to the fact that only the geodesic part of the curvature is an intrinsic property and hence it can be detected by a 2D inhabitant, therefore if this part of the curvature vanished the 2D inhabitant will fail to detect any curvature to the curve which is equivalent for him to having a straight line. Any deviation from such "straight lines" within the surface is therefore a geodesic curvature and hence it can be detected intrinsically by a 2D inhabitant.

Although a geodesic curve is frequently the curve of the shortest distance between two points on the surface it is not necessarily so. For instance, the largest of the two arcs forming a great circle on a sphere is a geodesic curve but it is not the curve of the shortest distance on the sphere between its two end points; in fact it is the curve of the longest distance among the circular arcs connecting the two points (Fig. 45). A similar example is the two arcs of a parallel circle on a circular cylinder connecting two points where the two arcs are different in length. Anyway, if on a surface $S$ there is exactly one geodesic curve connecting two given points, $P_1$ and $P_2$, then the length of the geodesic curve segment between $P_1$ and $P_2$ is the shortest distance on $S$ between these points.

## 5.7 Geodesic Curve

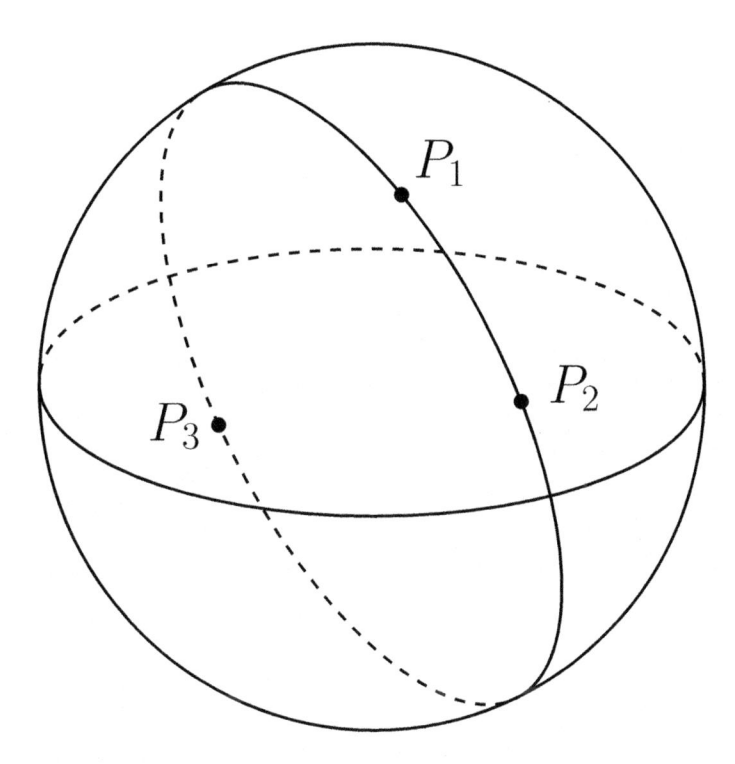

Figure 45: Two geodesic curves connecting two points, $P_1$ and $P_2$, on the surface of a sphere: a short one between $P_1$ and $P_2$ directly, and a long one between $P_1$ and $P_2$ through $P_3$. Both of these geodesics are arcs of a great circle on the sphere.

Based on the previous statements, being a shortest path is a sufficient but not necessary condition for being a geodesic, that is all shortest paths connecting two given points are geodesics but not all geodesics are shortest paths. A constraint may be imposed to make the criterion of minimal length apply to all geodesics by stating that geodesics minimize distance locally but not necessarily globally where an infinitesimal element of arc is considered in this constraint. Anyway, the universal criterion that should be adopted to identify geodesic curves is the vanishing of the geodesic curvature over the whole curve, as stated at the start of this section.

The geodesic, even in its restricted sense as the curve of the shortest distance, is not necessarily unique; for example all semi-circular meridians of longitude connecting the two poles (or in fact all semi-circular arcs connecting any two antipodal points) of a sphere are geodesics even in that sense and there is an infinite number of them. In fact, even the existence, not only uniqueness, of a geodesic connecting two points on a surface is not guaranteed. An example is the $xy$ plane excluding the origin of coordinates with two points on a straight line lying in the plane and passing through the origin where there is a lower limit for the length of any curve connecting the two points (Fig. 46). This limit is the straight line segment connecting the two points but this segment cannot be a geodesic on the plane since it includes the origin which is not on the plane. Any curve $C$ (other than the straight line segment) on the plane connecting the two points cannot be a curve of shortest length, and hence a geodesic, since there is always another curve on the plane

## 5.7 Geodesic Curve

connecting the two points which is shorter than $C$. In this context, we note that on a plane surface all geodesic curves are straight lines and hence of shortest length.

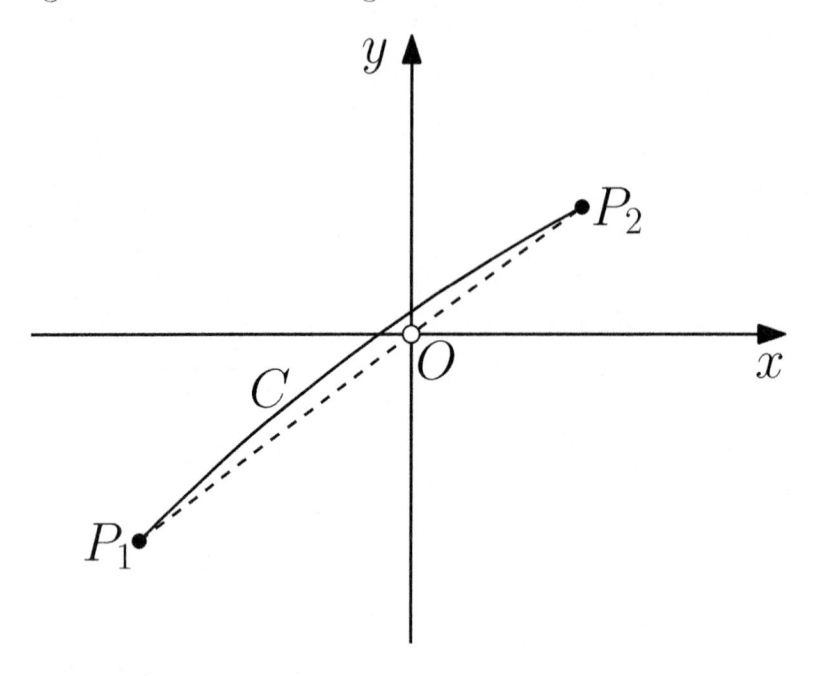

Figure 46: Non-existence of a geodesic curve connecting the two points $P_1$ and $P_2$ on the shown $xy$ plane which does not include the origin of coordinates $O$. The dashed line represents the straight line segment connecting $P_1$ and $P_2$ while the solid line $C$ represents other curves on the plane between the two points.

In the neighborhood of a given point $P$ on a surface and for any specific direction, there is exactly one geodesic curve passing through $P$ in that direction. More technically, for any specific point $P$ on a surface $S$ of class $C^3$, and for any tangent vector **v** in the tangent space of $S$ at $P$, there exists a geodesic curve on the surface in the direction of **v** that passes through $P$. In fact, this is based on the existence of a unique solution to the geodesic differential equations (Eqs. 416-417) when initial values of a point on the curve and its derivative (which represents its tangent direction) at that point are given. An obvious example of the previous statement is the plane where a straight line passes through any point and in any direction. Another example is the sphere where a great circle passes through any point and in any direction. A less obvious example is the cylinder where a helix (including the straight line generators and the circles which can be regarded as degenerate forms of helix) passes through any point and in any direction. Similarly, there is exactly one geodesic curve passing through two sufficiently close points on a smooth surface.[27]

As indicated before, geodesics in curved spaces are the equivalent of straight lines in flat spaces. For planes (or in fact for any Euclidean $n$D manifold) there *exists* a *unique*

---
[27] In this type of statement, which is found in common textbooks on differential geometry, we may need to add extra restrictions such as excluding closed surfaces or adding further conditions like "in the immediate neighborhood of the two points" to make the statement applicable to all types of surface.

## 5.7 Geodesic Curve

geodesic passing between any two points (whether the two points are close or not) which is the straight line segment connecting the two points.

The necessary and sufficient condition that should be satisfied by a naturally parameterized curve on a surface, both of class $C^2$, to be a geodesic curve is given by the following set of second order non-linear differential equations:

$$\frac{d^2 u^1}{ds^2} + \Gamma^1_{11} \left(\frac{du^1}{ds}\right)^2 + 2\Gamma^1_{12} \frac{du^1}{ds}\frac{du^2}{ds} + \Gamma^1_{22} \left(\frac{du^2}{ds}\right)^2 = 0 \qquad (416)$$

$$\frac{d^2 u^2}{ds^2} + \Gamma^2_{11} \left(\frac{du^1}{ds}\right)^2 + 2\Gamma^2_{12} \frac{du^1}{ds}\frac{du^2}{ds} + \Gamma^2_{22} \left(\frac{du^2}{ds}\right)^2 = 0 \qquad (417)$$

where $s$ is the arc length, and the Christoffel symbols are derived from the surface metric. The last equations can be merged in a single equation using tensor notation, that is:

$$\frac{\delta}{\delta s}\left(\frac{du^\alpha}{ds}\right) \equiv \frac{d^2 u^\alpha}{ds^2} + \Gamma^\alpha_{\beta\gamma} \frac{du^\beta}{ds}\frac{du^\gamma}{ds} = 0 \qquad (418)$$

where $\alpha, \beta, \gamma = 1, 2$ and the standard notation of absolute derivative is in use (see § 7). These equations, which can be obtained from Eq. 323 by setting the two components of the geodesic curvature vector to zero, have no closed form explicit solutions in general because of their non-linearity. Similar equations are used to identify the geodesic curves in general $n$D spaces.

From Eq. 418, it can be seen that being a geodesic is an intrinsic property since the conditions represented by this equation depend exclusively on the Christoffel symbols which depend only on the coefficients of the first fundamental form and their partial derivatives. Hence, geodesic curves can be detected and measured by a 2D inhabitant. From Eq. 418, it can also be seen that for planes (or indeed for any Euclidean $n$D manifold) the geodesic is a straight line since in this case the Christoffel symbols vanish identically and Eq. 418 will be reduced to $\frac{d^2 u^\alpha}{ds^2} = 0$ which has a straight line solution.

From Eq. 326, we see that the $u^1$ coordinate curves on a sufficiently smooth surface are geodesics *iff* $\Gamma^2_{11} = 0$. Similarly, from Eq. 328, we see that the $u^2$ coordinate curves are geodesics *iff* $\Gamma^1_{22} = 0$. We also see from Eqs. 327 and 329 that for coordinate systems with orthogonal coordinate curves, the coordinate curves are geodesics *iff* $E$ is independent of $v$ and $G$ is independent of $u$.

For a Monge patch of the form $\mathbf{r}(u, v) = (u, v, f(u, v))$, the geodesic differential equations are given by:

$$\left(1 + f_u^2 + f_v^2\right) u'' + f_u f_{uu}(u')^2 + 2 f_u f_{uv} u' v' + f_u f_{vv}(v')^2 = 0 \qquad (419)$$

$$\left(1 + f_u^2 + f_v^2\right) v'' + f_v f_{uu}(u')^2 + 2 f_v f_{uv} u' v' + f_v f_{vv}(v')^2 = 0 \qquad (420)$$

where the subscripts $u$ and $v$ represent partial derivatives of $f$ with respect to the surface coordinates $u$ and $v$, and the prime represents derivatives with respect to a natural parameter.

Based on what we have seen so far, it can be concluded that each one of the following provisions is a necessary and sufficient condition for a curve $C$ on a surface $S$ to be a geodesic curve:

## 5.7 Geodesic Curve

1. The geodesic component of the curvature vector is zero at each point on the curve, that is $\mathbf{K}_g = \mathbf{0}$ identically. This is based on the definition of geodesic curve which we stated earlier.
2. The osculating plane of the curve at each point of the curve is orthogonal to the tangent plane of $S$ at that point. The reason is that on geodesic curves $\kappa_g = 0$ and hence $\mathbf{n}$ and $\mathbf{N}$ are in the same orientation (parallel or anti-parallel) on all points along the curve (refer to Eqs. 305 and 307) and hence $\mathbf{n}$ lies in the osculating plane and the osculating plane will be orthogonal to the tangent plane.
3. The normal vector $\mathbf{n}$ to the surface at any point on the curve lies in the osculating plane. This is because for a geodesic curve, $\mathbf{K}_g$ vanishes identically and hence $\mathbf{K} = \kappa \mathbf{N} = \kappa_n \mathbf{n} = \mathbf{K}_n$.
4. The principal normal vector $\mathbf{N}$ of $C$ is normal to the surface at each point on $C$ since $\mathbf{N}$ is collinear with $\mathbf{n}$.
5. The curvature vector $\mathbf{K}$ of the curve is normal to the tangent plane of the surface at each point on the curve.

Being a geodesic is independent of the choice of the coordinate system and hence it is invariant under permissible transformations. It is also independent of the type of representation and parameterization and hence it is invariant in this sense.

Geodesic curves can be open or closed curves and may be self-intersecting. In this statement, we are considering the totality of the geodesic path as characterized by having identically vanishing geodesic curvature and not as a connecting arc between two distinct points.[28] Examples of open geodesics are the straight lines on planes and the helices on cylinders while examples of closed geodesics are the geodesics of spheres which are great circles and the parallel circles on cylinders. In fact, all the geodesics on sphere are closed curves as they are great circles, while circles on cylinder is the only case of closed geodesics on this type of surface.

As a result of the Gauss-Bonnet theorem (see § 4.8), on a surface with negative Gaussian curvature two geodesics cannot intersect at more than one point if the geodesics enclose a simply-connected region. The reason is that on introducing an artificial vertex at a regular point on one of these curves we will have a new corner with $\pi$ interior angle and hence the sum of the angles of the geodesic triangle will exceed $\pi$ which is impossible on a surface with $K < 0$. Also, on introducing an artificial vertex at a regular point on each one of these curves we will have a geodesic quadrilateral whose internal angles add up to more than $2\pi$ on a surface with $K < 0$ which is not possible (see § 4.8).

Another result of the Gauss-Bonnet theorem is that a surface with negative Gaussian curvature cannot have a geodesic that intersects itself. This may be established by a similar argument to the previous one that is: on introducing two artificial corners at two regular points on the curve, we will have a geodesic triangle whose interior angles add up to more than $\pi$ which is not possible on a surface with $K < 0$ (see § 4.8).

On a patch of a surface of class $C^2$ with orthogonal coordinate curves and with the

---

[28] To put it in a different way, "geodesic curves" has two common uses: (a) curves with identically vanishing geodesic curvature and (b) arcs of optimal length connecting two points, where (b) in a sense is a subset of (a). Here, "geodesic curves" is used in the first sense.

first fundamental form coefficients being dependent on only the $u$ coordinate variable (i.e. $E = E(u)$, $F = 0$ and $G = G(u)$) the following statements apply:
1. The $u$ coordinate curves are geodesics.
2. The $v$ coordinate curves are geodesics *iff* $\partial_u G = 0$ along these curves.
3. A curve $C$ represented by $\mathbf{r} = \mathbf{r}(u, v(u))$ is a geodesic *iff*:

$$v = \pm \int_C \frac{k\sqrt{E}}{\sqrt{G(G - k^2)}} du \tag{421}$$

where $k$ is a constant.

The case of dependence on only the $v$ coordinate variable can be obtained by re-labeling the coordinate variables and coefficients. The second of the above statements may be generalized by saying: on a surface with orthogonal coordinate curves, the curves of constant $u^\alpha$ are geodesics *iff* $a_{\beta\beta}$ ($\beta \neq \alpha$) is a function of $u^\beta$ only.

As indicated before, geodesics in curved spaces represent a generalization of straight lines in flat spaces. Hence, geodesics may be described as the straightest curves in the space. In fact, geodesic curves on a developable surface become straight lines when the surface is developed into a plane by unrolling. This may be demonstrated by the perception of a 2D inhabitant of the surface who will fail to observe any difference to the geodesic curve when the surface is developed into a plane and the geodesic curve necessarily becomes a straight line on the plane. More generally, a geodesic curve will be mapped onto a geodesic curve by any isometric transformation due to the invariance of the geodesic curvature under this type of transformations since geodesic curvature is an intrinsic property. Therefore, isometric surfaces possess identical geodesic equations.

Another sufficient and necessary condition for a surface curve to be geodesic is being a tangent to a parallel vector field. A vector attained by parallel propagation (see § 2.7) of a tangent vector to a geodesic curve stays always tangent to the geodesic curve. As a result, a vector field attained by parallel propagation along a geodesic makes a constant angle with the geodesic.

## 5.8 Line of Curvature

A "line of curvature" is a curve $C$ on a surface $S$ defined on an interval $I \subseteq \mathbb{R}$ as $C: I \to S$ with the condition that the tangent of $C$ at each point on $C$ is collinear with one of the principal directions (see § 4.4) of the surface at that point. We note that "line" here does not mean straight. Since the definition of the line of curvature is seemingly based on the existence of distinct principal directions, umbilical points (see § 4.10) may be excluded from the above definition of the line of curvature due to the absence of distinct principal directions at these points although there seems to be no harm in including isolated umbilical points (at least) over the path of the line of curvature.[29] Referring to Eq. 134,

---

[29] There are some details about this issue that could be elaborated where different conventions should be considered. The main issue is that: can the number of principal directions at a given point on a surface exceed two or not. Hence, some of the future materials may not be based on a single convention.

## 5.8 Line of Curvature

on a line of curvature either $\sin\theta = 0$ or $\cos\theta = 0$ and hence the lines of curvature are characterized by having identically vanishing geodesic torsion (i.e. $\tau_g = 0$).

The condition that should be satisfied by a line of curvature is usually given by the following relation:

$$(a_{12}b_{11} - a_{11}b_{12})\,du^1 du^1 + (a_{22}b_{11} - a_{11}b_{22})\,du^1 du^2 + (a_{22}b_{12} - a_{12}b_{22})\,du^2 du^2 = 0 \quad (422)$$

where the indexed $a$ and $b$ are the coefficients of the surface covariant metric and covariant curvature tensors respectively. In fact, this is the same as the condition given by Eq. 350 for the principal directions, which is consistent with the fact that the line of curvature is aligned along a principal direction at each of its points. The condition that should be satisfied by a line of curvature on a surface may be given in tensor notation by:

$$\underline{\epsilon}^{\gamma\delta} a_{\alpha\gamma} b_{\beta\delta} du^\alpha du^\beta = 0 \quad (423)$$

where $\underline{\epsilon}^{\gamma\delta}$ is the 2D absolute permutation tensor.

Examples of lines of curvature are meridians and parallels of surface of revolution of class $C^2$. For a developable surface, the lines of curvature consist of its generators and their orthogonal trajectories. On a sufficiently smooth surface, any geodesic which is a plane curve is a line of curvature. Similarly, on a sufficiently smooth surface, if a geodesic curve $C$ is a line of curvature then $C$ is a plane curve. The lines of intersection of each pair of a triply orthogonal system are also lines of curvature. We remark that three families of surfaces in a subset $V$ of a 3D space form a triply orthogonal system if at each point $P$ of $V$ there is a single surface of each family passing through $P$ such that each pair of these surfaces intersect orthogonally at their curve of intersection.

At a non-umbilical point $P$ on a sufficiently smooth surface $S$, the $u^1$ and $u^2$ coordinate curves are aligned with the principal directions *iff* $f = F = 0$ at $P$. The "if" part can be seen, for example, from Eq. 350 which in this case (i.e. $f = F = 0$) will reduce to $(gE - eG)\,du^1 du^2 = 0$ and hence it will be satisfied on the coordinate curves, since $du^2$ will vanish on the $u^1$ coordinate curve while $du^1$ will vanish on the $u^2$ coordinate curve, and these curves become aligned with the principal directions. The "only if" part can also be seen from Eq. 350 because if the coordinate curves are aligned with the principal directions then this equation should be satisfied where it becomes $(fE - eF)\,du^1 du^1 = 0$ for the $u^1$ coordinate curve and $(gF - fG)\,du^2 du^2 = 0$ for the $u^2$ coordinate curve and both of these equations imply $f = F = 0$ since $du^1 \neq 0$ on the $u^1$ coordinate curve and $du^2 \neq 0$ on the $u^2$ coordinate curve. We note that by considering the stated conditions and the definitions of the coefficients of the first and second fundamental forms as well as the Rodrigues curvature formula (see § 4.4), it can be concluded that the coefficients $E, e$ and $g, G$ cannot vanish.

As a consequence of the last paragraph, the coordinate curves on the surface $S$, excluding the umbilical points, are lines of curvature *iff* $f = F = 0$ over the entire surface. This may also be stated by saying that on a smooth surface, excluding planes and spheres (whose all points are umbilical), if the lines of curvature are selected as the net of coordinate curves then $a_{12} = b_{12} = 0$ over the entire surface excluding the umbilical points. We remark that

## 5.9 Asymptotic Line

when the $u$ and $v$ coordinate curves of a surface patch are lines of curvature, the principal curvatures, $\kappa_1$ and $\kappa_2$, over the entire patch will be given by:

$$\kappa_1 = \frac{e}{E} \qquad \kappa_2 = \frac{g}{G} \qquad (424)$$

where $E, G, e, g$ are the coefficients of the first and second fundamental forms at the points of the patch. The reader is referred to § 4.4 for justification (see Eq. 354 and related text).

On a surface of class $C^3$, there are two perpendicular families of lines of curvature in the neighborhood of any non-umbilical point. If the curve of intersection of two surfaces is a line of curvature for one surface then it is a line of curvature for the other surface when the two surfaces are intersecting each other at a constant angle.

The lines of curvature form a real orthogonal grid over the surface. A curve is a line of curvature *iff* the tangent to the curve and the tangent to its spherical image (see § 3.10) at their corresponding points are parallel. The lines of curvature on a surface, which is not a sphere or minimal surface (see § 6.7), are represented by an orthogonal net on its spherical image.

As indicated above, in the neighborhood of a non-umbilical point on a sufficiently smooth surface there are two orthogonal families of lines of curvature. Hence, at each point $P$ on such a surface a coordinate patch including $P$ can be introduced in the neighborhood of $P$ where the coordinate curves at $P$ are aligned with the principal directions. On a surface patch where the Gaussian curvature does not vanish, the angles between the asymptotic lines (see § 5.9) are bisected by the lines of curvature.

## 5.9 Asymptotic Line

An asymptotic direction of a surface at a given point $P$ is a direction for which the normal curvature vanishes, i.e. $\kappa_n = 0$. Hence, in an asymptotic direction at a point on a surface we have (see Eq. 307):

$$\mathbf{K} = \mathbf{K}_g = \kappa_g \mathbf{u} \qquad (425)$$

As a consequence of Eq. 316, $\kappa_n$ is zero in the directions for which the second fundamental form is zero. Hence, the necessary and sufficient condition for the asymptotic directions is that:

$$b_{\alpha\beta} du^\alpha du^\beta = b_{11}(du^1)^2 + 2b_{12} du^1 du^2 + b_{22}(du^2)^2 = 0 \qquad (426)$$

We note that asymptotic directions are defined only at points for which the Gaussian curvature is non-positive ($K \leq 0$) and hence it is not defined at elliptic points (see § 4.9) where $K > 0$. This is because at elliptic points either $\kappa_1 > 0$ and $\kappa_2 > 0$ or $\kappa_1 < 0$ and $\kappa_2 < 0$ and hence $\kappa_n$ cannot take the value zero at these points.

As a result, the number of asymptotic directions at elliptic, parabolic and hyperbolic points is 0, 1 and 2 respectively, while at flat points all directions are asymptotic. The two asymptotic directions of a hyperbolic point separate the directions of positive normal curvature from the directions of negative normal curvature. The sign of the normal curvature at elliptic and parabolic points is the same in all directions, excluding the asymptotic

## 5.9 Asymptotic Line

direction of the parabolic point. Similarly, at flat points the normal curvature is zero in all directions.

A $t$-parameterized surface curve $C(t) : I \to S$, where $I \subseteq \mathbb{R}$ is an open interval and $S$ represents the surface, is described as an asymptotic line if at each point $t \in I$ the vector $\mathbf{T}$, which is the tangent to the curve, is collinear with an asymptotic direction at that point. It should be remarked that "line" here is not required to be straight; hence asymptotic lines are also called asymptotic curves.

From the above statements, it can be seen that the asymptotic lines are characterized by the following features:

1. The normal component of the curvature vector is zero at each point on the curve, that is $\mathbf{K}_n = \mathbf{0}$ identically. This is based on the definition of asymptotic line as stated above.
2. The tangent plane to the surface at each point of the curve coincides with the osculating plane of the curve at that point. This is a consequence of having identically vanishing normal curvature, since the curvature vector will then have only a tangential component and hence the osculating plane at each point of an asymptotic line becomes tangent to the surface at that point.

The differential equation representing asymptotic lines can be obtained from the condition that the normal curvature vanishes identically over the line, that is:

$$e \left( \frac{du^1}{ds} \right)^2 + 2f \frac{du^1}{ds} \frac{du^2}{ds} + g \left( \frac{du^2}{ds} \right)^2 = 0 \qquad (427)$$

which is based on Eqs. 316 and 233 or on Eq. 318. The necessary and sufficient condition for the $u^1$ and $u^2$ coordinate curves to become asymptotic lines is that $e = 0$ identically on the $u^1$ coordinate curve and $g = 0$ identically on the $u^2$ coordinate curve.[30] This is based on Eqs. 319 and 320 which are fully justified there.

According to Eq. 308, $\kappa_n = \mathbf{n} \cdot \mathbf{K}$ where $\mathbf{n}$ and $\mathbf{K}$ are respectively the normal vector to the surface and the curve curvature vector. Hence, a curve on a sufficiently smooth surface is an asymptotic line *iff* $\mathbf{n} \cdot \mathbf{K} = 0$ identically. Now, since the vector $\mathbf{n}$ cannot vanish on the regular points of the surface, then this condition is realized if at each point on the curve either $\mathbf{K} = \mathbf{0}$ or $\mathbf{K}$ and $\mathbf{n}$ are orthogonal vectors. In the former case the point is an inflection point while in the latter case the osculating plane is tangent to the surface at the point. Therefore, all points on an asymptotic line should be one of these types or the other. The reverse is also true, i.e. a curve whose all points are one of these types or the other is an asymptotic line. As a result of the last statements, any straight line on a surface is an asymptotic line since the curve curvature vector $\mathbf{K}$ vanishes identically on such a line.

According to the theorem of Beltrami-Enneper, along an asymptotic non-straight line on a sufficiently smooth surface the square of the torsion $\tau$ is equal to the negative of the Gaussian curvature $K$, that is:

$$\tau^2 = -K \qquad (428)$$

---

[30] These conditions can be taken together or separately.

## 5.9 Asymptotic Line

where $\tau$ and $K$ are evaluated at each individual point along the curve. Since asymptotic directions are defined only at points for which $K \leq 0$, the square of the torsion in the above equation is equal to the absolute value of the Gaussian curvature at the point, that is: $\tau^2 = |K|$ and hence $\tau$ is real as it should be. The torsions of two asymptotic lines passing through a given point on a sufficiently smooth surface are equal in magnitude and opposite in sign.

As we will see (refer to § 5.10), asymptotic directions are self-conjugate. In fact, some authors take self-conjugation as the defining characteristic for being asymptotic. From the definition of the asymptotic direction plus the Euler equation (Eq. 341), we see that the angle $\theta$ which an asymptotic direction makes with the principal direction of $\kappa_1$ at a given non-umbilical point $P$ on a sufficiently smooth surface $S$ is given by:

$$\tan^2 \theta = -\frac{\kappa_1}{\kappa_2} \qquad (429)$$

where $\kappa_1$ and $\kappa_2$ are the principal curvatures of $S$ at $P$. When $\kappa_2 = 0$, the reciprocal of this relation should be taken. Again, $\tan \theta$ is real since at points with $K < 0$, the two kappas should have opposite signs (see Eq. 355). The situation is similar when $\kappa_1 = 0$ and $\kappa_2 < 0$. However, when $\kappa_1 > 0$ and $\kappa_2 = 0$ the reciprocal will be taken and the cotangent will be real. The possibility of $\kappa_1 = \kappa_2 = 0$ is already excluded by the non-umbilical condition since a point with $\kappa_1 = \kappa_2 = 0$ is a flat umbilic.

Because Eq. 426 is quadratic, it possesses two solutions which are real and distinct, or real and coincident, or conjugate imaginary depending on its discriminant $\Delta$ which is opposite in sign to the determinant $b$ of the surface covariant curvature tensor.[31] Hence, the asymptotic directions at a given point on a surface can be classified according to the determinant $b$ at the point as:

1. Real and distinct for $\Delta > 0$ and hence $b < 0$.
2. Real and coincident for $\Delta = 0$ and hence $b = 0$.
3. Conjugate imaginary for $\Delta < 0$ and hence $b > 0$.

This is inline with the above statement that the number of asymptotic directions at elliptic, parabolic and hyperbolic points is 0, 1 and 2 respectively because, as seen in § 4.9, the local shape of a surface at a given point is determined by the sign of $b$ at the point where $b < 0$ at hyperbolic point, $b = 0$ at parabolic and flat points, and $b > 0$ at elliptic point. We also remark that from Eq. 356 we can see that the sign of the Gaussian curvature $K$ is the same as the sign of $b$ due to the fact that $a > 0$ since the first fundamental form is positive definite, as established earlier. Hence, the above-described classification of the asymptotic directions can also be based on $K$, as stated for $b$ in the previous points. Again, as seen in § 4.9, the sign of $K$ is used to determine the local shape at a point, and hence the number of the asymptotic directions is determined accordingly.

It is worth noting that on a sufficiently smooth surface with orthogonal families of asymptotic lines the mean curvature $H$ is zero. This can be seen from Eq. 383 where by aligning the coordinate curves along the asymptotic directions with a proper labeling, we will get:

---

[31] The discriminant is: $\Delta = 4(b_{12})^2 - 4b_{11}b_{22}$ while the determinant is: $b = b_{11}b_{22} - (b_{12})^2$, and hence $\Delta = -4b$.

$F = 0$ since the coordinate curves are orthogonal, and $e = g = 0$ according to the above condition which we stated after Eq. 427, and hence $H = 0$. Now since $H$ is invariant, then this will remain valid under permissible transformations to other surface coordinates. Another note is that the principal directions at a given point on a smooth surface bisect the asymptotic directions at the point, as indicated before. Also, on a smooth surface of class $C^3$, there are two distinct families of asymptotic directions in the neighborhood of any hyperbolic point.

As seen before, any straight line contained in a surface is an asymptotic line. As well as the previously stated explanation, this may also be justified by the fact that such a line is wholly contained in a plane, which is the tangent space of each of its points, and hence the line is an asymptotic line with an identically vanishing normal curvature.

## 5.10 Conjugate Direction

A direction $\frac{\delta v}{\delta u}$ at a point on a sufficiently smooth surface is described as conjugate to the direction $\frac{dv}{du}$ iff the following relation holds true:[32]

$$d\mathbf{r} \cdot \delta\mathbf{n} = 0 \qquad (430)$$

where:

$$d\mathbf{r} = \mathbf{E}_1 du + \mathbf{E}_2 dv \qquad \delta\mathbf{n} = \partial_u \mathbf{n}\, \delta u + \partial_v \mathbf{n}\, \delta v \qquad (431)$$

From Eq. 244, it can be seen that the condition given by Eq. 430 is equivalent to the following condition:

$$e\, du\delta u + f\,(du\delta v + dv\delta u) + g\, dv\delta v = 0 \qquad (432)$$

The last equation may also be obtained directly by substituting from Eq. 431 into Eq. 430 and performing the dot product with the use of Eqs. 249-251 to obtain the coefficients $e, f, g$. Due to the symmetry in the above relations, $\frac{dv}{du}$ is also conjugate to $\frac{\delta v}{\delta u}$, and hence the two directions are described as conjugate directions. At a hyperbolic or an elliptic point on a sufficiently smooth surface, each direction has a unique conjugate direction. As stated earlier, an asymptotic direction is a self-conjugate direction.

Two families of curves on a sufficiently smooth surface are described as conjugate families if the directions of their tangents at each intersection point of the curves are conjugate directions. The $u$ and $v$ coordinate curves on a smooth surface are conjugate families of curves *iff* $f$, which is the coefficient of the second fundamental form, vanishes identically. This can be seen from Eq. 432 where by a proper labeling of the $u$ and $v$ coordinates in the two directions to make $du\delta u = dv\delta v = 0$ on the coordinate curves the first and last terms of Eq. 432 will vanish on the coordinate curves and hence the curves will be conjugate families by satisfying the reduced condition $f(du\delta v + dv\delta u) = 0$ which leads to the condition $f = 0$ since $du\delta v + dv\delta u \neq 0$ according to this labeling. Similarly, if $f = 0$ then the coordinate curves will satisfy the reduced condition $f(du\delta v + dv\delta u) = 0$ and hence they are conjugate families.

---

[32] The notation $\frac{\delta v}{\delta u}$ is not related to the notation of absolute derivative (see § 7).

## 5.11 Exercises

5.1 State two criteria for a space curve to be straight.

5.2 Prove that a curve represented by $\mathbf{r}(t)$ is a straight line if $\dot{\mathbf{r}}$ and $\ddot{\mathbf{r}}$ are linearly dependent over the whole curve.

5.3 Show that a space curve whose all tangent lines are parallel is a straight line.

5.4 Correct, if necessary, the following statement: "All straight lines on a surface are geodesic curves and vice versa".

5.5 What is the characteristic feature of plane curves? From this, explain why the torsion of plane curves is identically zero.

5.6 Prove that a curve is a plane curve if its osculating planes have a common intersection point.

5.7 Show that a space curve represented by $\mathbf{r}(t)$ is a plane curve *iff* $\dot{\mathbf{r}} \cdot (\ddot{\mathbf{r}} \times \dddot{\mathbf{r}})$ vanishes identically.

5.8 Prove that having an identically vanishing torsion is a necessary and sufficient condition for a curve to be a plane curve.

5.9 Show that two curves are plane curves if they have the same binormal lines at each pair of their corresponding points.

5.10 Define, rigorously, involute and evolute curves making a simple plot to outline their relation. Also explain the role of the tangent surface of the evolute in this context.

5.11 Explain all the symbols used in the following equation which is related to involute curves: $\mathbf{r}_i = \mathbf{r}_e + (c - s)\,\mathbf{T}_e$. Make sense of this equation using your plot in the previous exercise.

5.12 Outline the visual demonstration which is commonly used to explain the relation between an involute and its evolute. Use the plot mentioned in the last two questions in your explanation.

5.13 Show that the tangent line of a curve and the principal normal line of its evolute are parallel at their corresponding points.

5.14 Prove that the evolutes of plane curves are helices.

5.15 Prove that for a plane curve $C$ the locus of the centers of curvature of $C$ is an evolute of $C$.

5.16 Derive the parametric equation of the involute of a circle represented by: $\mathbf{r}(\theta) = (5\cos\theta, 5\sin\theta)$ where $0 \leq \theta < 2\pi$.

5.17 Prove that any two involutes of a plane curve are associated Bertrand curves.

5.18 How many involutes a given curve can have? How these involutes are related to each other through the constant $c$ (see Eq. 407)?

5.19 How many evolutes a given curve can have? How these evolutes are related to each other through the constant $c$ (see Eq. 407)?

5.20 Justify the fact that the involutes of a circle are congruent with a clear explanation of how these involutes are related to each other.

5.21 Define Bertrand curves outlining two of their main characteristic features.

5.22 Show that a helix has an infinite number of Bertrand associates and identify these associates.

## 5.11 Exercises

5.23 Prove that on a pair of Bertrand curves, the angle between their tangents at corresponding points is constant.

5.24 State a sufficient and necessary condition for a curve to be a Bertrand curve by having an associate Bertrand curve.

5.25 Show that a plane curve has always a Bertrand associate.

5.26 Prove that the product of torsions at the corresponding points of a pair of associated Bertrand curves is constant (see Eq. 408 and related text for explanation).

5.27 Prove that on a pair of Bertrand curves, the distance between their corresponding points is constant.

5.28 Give a brief definition of spherical indicatrix with a simple sketch of the spherical normal indicatrix $\bar{C}_\mathbf{N}$ of a space curve to illustrate this concept.

5.29 Prove the following equation (Eq. 410): $\kappa_\mathbf{T}^2 = \frac{\kappa^2 + \tau^2}{\kappa^2}$.

5.30 Discuss the similarities and differences between Gauss mapping (see § 3.10) and spherical indicatrix mapping.

5.31 Justify, using a simple fact about helices, that the spherical images of $\mathbf{T}, \mathbf{N}, \mathbf{B}$ of a helix rotating around the $z$-axis are circles centered around the $z$-axis.

5.32 Justify the following statement: "The binormal indicatrix is a single point for a plane curve, and the tangent indicatrix is a single point for a straight line".

5.33 Prove, rigorously, that the tangent indicatrix of a helix is a circle.

5.34 Prove that the tangent to the spherical indicatrix of the tangent to a given space curve $C$ and the principal normal of $C$ are parallel.

5.35 Write down the mathematical formula for the torsion of the spherical binormal indicatrix of a space curve explaining all the symbols used in the formula.

5.36 What "spherical curve" means? give a common example of a spherical curve.

5.37 State, mathematically, the sufficient and necessary condition for a curve to be a spherical curve explaining all the symbols involved.

5.38 Investigate if the curve represented parametrically by: $\mathbf{r}(t) = (5\cos t, 5\cos t \sin t, 5\sin^2 t)$ is a spherical curve or not.

5.39 Show that the spherical image of a curve $C(t)$ is a closed curve when the vector that generates the spherical image is a periodic function of $t$ although $C$ may not be periodic. Discuss in this context the helix as an example.

5.40 What is the characteristic feature of geodesic curves?

5.41 Give a rigorous mathematical definition of geodesic curve.

5.42 Give examples of geodesic curves on the following surfaces: plane, sphere and cylinder.

5.43 On a surface of revolution, what type of curve is necessarily geodesic and what type is potentially geodesic?

5.44 Define geodesic curve variationally using the concepts of calculus of variations.

5.45 Show that any helix on a circular cylinder is a geodesic curve.

5.46 Outline the relation between the concept of geodesic curve and the concept of curve of shortest distance between two points.

5.47 Prove that Eq. 418 is a sufficient and necessary condition for a curve to be geodesic.

5.48 Find an analytical expression representing the geodesic curves on a circular cone using one of its parametric representations.

5.49 Outline the concept of geodesic curve on a surface as perceived by a 2D inhabitant of the surface.

5.50 Prove that all the geodesic curves on a plane are straight lines.

5.51 Does a geodesic curve necessarily exist between two given points on a space surface, and if it does exist is the geodesic curve necessarily unique? Support your answer with illustrating examples for both cases.

5.52 Discuss the following statement and its implications: "Being a shortest path is a sufficient but not necessary condition for being a geodesic curve".

5.53 Correct, if necessary, the following statement: "In the neighborhood of a given point $P$ on a surface, there is exactly one geodesic curve that passes through $P$".

5.54 Write down, with full explanation, the differential equations which provide the necessary and sufficient conditions for a naturally parameterized curve on a surface to be geodesic.

5.55 Correct the following relations which represent the geodesic differential equations for a Monge patch of the form $\mathbf{r}(u,v) = (u, v, f(u,v))$:

$$\left(1 + f_u^2 + f_v^2\right) u' + f_u f_{uu}(u')^2 + 2 f_u f_{uv} u' v' + f_u f_{vv}(v')^2 = 0$$
$$\left(1 + f_u^2 + f_v^2\right) v'' + f_v f_u(u')^2 + 2 f_v f_{uv} u' v' + f_v f_{vv}(v')^2 = 0$$

5.56 Why the normal vector to the surface at any point on a geodesic curve should be contained in the osculating plane of the curve at that point? Give a clear technical justification.

5.57 Using Gauss-Bonnet theorem, explain why a surface with negative Gaussian curvature cannot have a geodesic that intersects itself.

5.58 Using Eq. 418, prove that all meridians of a surface of revolution are geodesic curves.

5.59 Discuss the following statement in the context of the perception of geodesic curves by a 2D inhabitant: "Geodesic curves on a developable surface become straight lines when the surface is developed into a plane".

5.60 Give an example of a surface curve whose normal curvature and geodesic curvature are identically zero over the whole curve.

5.61 What is the relation between a line of curvature on a surface and the principal directions at the points of the curve?

5.62 Prove that if a plane and a surface are intersecting at a constant angle then their curve of intersection is a line of curvature.

5.63 Repeat the previous exercise replacing plane with sphere.

5.64 Can a line of curvature include umbilical points? Discuss this issue considering the question of allowing more than two principal directions at a point or not.

5.65 Prove that for any sufficiently smooth surface of revolution, the parallels and meridians are lines of curvature.

5.66 Show that for a given non-umbilical point $P$ on a sufficiently smooth surface there is a patch that contains $P$ where the directions of the coordinate curves at $P$ are principal directions.

5.67 Prove Hilbert lemma using the proposal that the coordinate curves on a patch can coincide with the lines of curvature in the neighborhood of a non-umbilical point.

5.11 Exercises

5.68 Prove that if the curve of intersection of two surfaces is a line of curvature for one surface then it is a line of curvature for the other surface when the two surfaces are intersecting each other at a constant angle.

5.69 Outline the role of geodesic torsion in characterizing the line of curvature employing a mathematical formulation in this context.

5.70 Give two examples for the line of curvature on specific types of surface discussing in each case why the described curve should be a line of curvature.

5.71 Give the formulae for the principal curvatures, $\kappa_1$ and $\kappa_2$, when the $u^1$ and $u^2$ coordinate curves of a surface patch are lines of curvature.

5.72 Using tensor notation, state the mathematical condition that should be met by a line of curvature on a space surface with full explanations of all the symbols involved.

5.73 Which types of surface should be excluded from the following statement: "The lines of curvature on a surface are represented by an orthogonal net on its spherical image"?

5.74 Prove that on a Monge patch of the form $\mathbf{r}(u,v) = (u, v, f(u,v))$, the coordinate curves are orthogonal family iff $f_u f_v = 0$ identically.

5.75 Give a mathematical condition for a direction on a surface at a given point to be asymptotic.

5.76 Prove that the asymptotic directions are bisected by the lines of curvature.

5.77 Give a rigorous technical definition of asymptotic line.

5.78 Why the second fundamental form at a point of a surface should vanish in the asymptotic direction?

5.79 Prove Beltrami-Enneper theorem (see Eq. 428 and surrounding text).

5.80 One of the characteristic features of asymptotic line is that the tangent plane to the surface at each point of the line coincides with the osculating plane of the line at that point. Why?

5.81 Show that on a smooth surface with orthogonal families of asymptotic lines the mean curvature is zero.

5.82 Justify the following statement: "The necessary and sufficient condition for the $u^1$ and $u^2$ coordinate curves to be asymptotic lines is that $e = 0$ identically on the $u^1$ coordinate curves and $g = 0$ identically on the $u^2$ coordinate curves".

5.83 Using Eq. 426, prove that the generators of a circular cylinder are asymptotic lines.

5.84 According to the theorem of Beltrami-Enneper we have: $\tau^2 = -K$ where $\tau$ and $K$ stand for torsion and Gaussian curvature. Does this mean that $\tau$ is imaginary? Fully justify your answer.

5.85 The angle $\theta$ which an asymptotic direction makes with the principal direction of $\kappa_1$ is given by: $\tan^2 \theta = -\frac{\kappa_1}{\kappa_2}$ where $\kappa_1$ and $\kappa_2$ are the principal curvatures. Derive this equation.

5.86 Classify the asymptotic directions at a given point on a surface as real and distinct, or real and coincident, or conjugate imaginary according to the determinant of the surface covariant curvature tensor at that point.

5.87 Why the classification in the previous question can also be based on the Gaussian curvature of the point?

5.88 The classification in the two previous questions is related to the number of asymptotic

## 5.11 Exercises

directions at elliptic, hyperbolic and parabolic points. How?

5.89 Justify the following statement: "A straight line contained in a surface is an asymptotic line".

5.90 State a mathematical condition for two directions at a given point on a surface to be conjugate directions explaining all the symbols used.

5.91 What "conjugate families of curves on a surface" means?

5.92 Show that at hyperbolic and elliptic points each direction has a unique conjugate direction.

5.93 Show that on a surface represented by $\mathbf{r} = \mathbf{r}_1(u) + \mathbf{r}_2(v)$, the coordinate curves are conjugate families of curves.

5.94 What is the necessary and sufficient condition for the $u^1$ and $u^2$ coordinate curves on a smooth surface to be conjugate families?

# Chapter 6
# Special Surfaces

There are many classifications to surfaces in 3D spaces depending on their properties and relations. A few of these classifications are briefly investigated in the following sections of this chapter.

## 6.1 Plane Surface

Planes are simple, ruled, connected, elementary surfaces. The following statements apply to planes:
1. All the coefficients of the surface curvature tensor vanish identically throughout plane surfaces.
2. The Riemann-Christoffel curvature tensor vanishes identically over plane surfaces.
3. The Gaussian curvature $K$ and the mean curvature $H$ vanish identically over planes.
4. Planes are minimal surfaces (see § 6.7).
5. All points on planes are flat umbilical.
6. At any point on a plane surface, $\kappa_1 = \kappa_2 = 0$ and hence all the directions are principal directions (or there is no principal direction).
7. At any point on a plane surface, all the directions are asymptotic.
8. A sufficient and necessary condition for a surface to be plane is having an identically vanishing surface curvature tensor.
9. A sufficient and necessary condition for a surface to be isometric with the plane is having an identically vanishing Riemann-Christoffel curvature tensor. The same applies for identically vanishing Gaussian curvature.

## 6.2 Quadratic Surface

Quadratic surfaces are defined by the following quadratic equation:

$$A_{ij}x^i x^j + B_i x^i + C = 0 \qquad (i,j = 1,2,3) \qquad (433)$$

where the coefficients $A_{ij}$ and $B_i$ are real-valued tensors of rank-2 and rank-1 respectively and $C$ is a real scalar. There are many degenerate and non-degenerate types of quadratic surface. However, we consider here only six non-degenerate types which are probably the most commonly occurring in differential geometry. These six types are: ellipsoid, hyperboloid of one sheet, hyperboloid of two sheets, elliptic paraboloid, hyperbolic paraboloid, and quadric cone. The first five of these quadratic surfaces have been defined in § 1.4.1 using parametric forms.

By rigid motion transformations, consisting of translation and rotation of coordinate system, whose purpose is to put the center of symmetry or vertex of these surfaces at the origin of coordinates and orient their axes and planes of symmetry with the coordinate lines and coordinate planes, these types can be given in the following canonical forms where we assume a Euclidean 3D space with a rectangular Cartesian coordinate system:

1. Ellipsoid (Fig. 47 a):
$$\frac{x^2}{a^2} + \frac{y^2}{b^2} + \frac{z^2}{c^2} = 1 \tag{434}$$

2. Hyperboloid of one sheet (Fig. 47 b):
$$\frac{x^2}{a^2} + \frac{y^2}{b^2} - \frac{z^2}{c^2} = 1 \tag{435}$$

3. Hyperboloid of two sheets (Fig. 47 c):
$$\frac{x^2}{a^2} - \frac{y^2}{b^2} - \frac{z^2}{c^2} = 1 \tag{436}$$

4. Elliptic paraboloid (Fig. 47 d):
$$\frac{x^2}{a^2} + \frac{y^2}{b^2} - z = 0 \tag{437}$$

5. Hyperbolic paraboloid (Fig. 47 e):
$$\frac{x^2}{a^2} - \frac{y^2}{b^2} - z = 0 \tag{438}$$

6. Quadric cone (Fig. 47 f):
$$\frac{x^2}{a^2} + \frac{y^2}{b^2} - \frac{z^2}{c^2} = 0 \tag{439}$$

We note that in the above equations $a, b, c$ are real parameters,[33] and for convenience we use $x, y, z$ for $x^1, x^2, x^3$ respectively. Also, for the first three of these surfaces the origin of coordinates is not a valid surface point, as seen from their equations.

## 6.3 Ruled Surface

A "ruled surface", or "scroll", is a surface generated by a continuous translational-rotational motion of a straight line in space. Hence, at each point of the surface there is a straight line passing through the point and lying entirely in the surface. Planes, cones, cylinders and Mobius strips (Fig. 1) are common examples of ruled surface. The parabolic cylinder (Fig. 48) is another example of ruled surface. The different perspectives of the generating line along its movement are described as the rulings of the surface. A ruled surface that can be generated by two different families of lines is called doubly-ruled surface. Examples of doubly-ruled surface are hyperbolic paraboloids (Fig. 49) and hyperboloids of one sheet (Fig. 50). At any point of a regular ruled surface, the Gaussian curvature is non-positive ($K \leq 0$).

## 6.3 Ruled Surface

(a) Ellipsoid

(b) Hyperboloid of one sheet

(c) Hyperboloid of two sheets

(d) Elliptic paraboloid

(e) Hyperbolic paraboloid

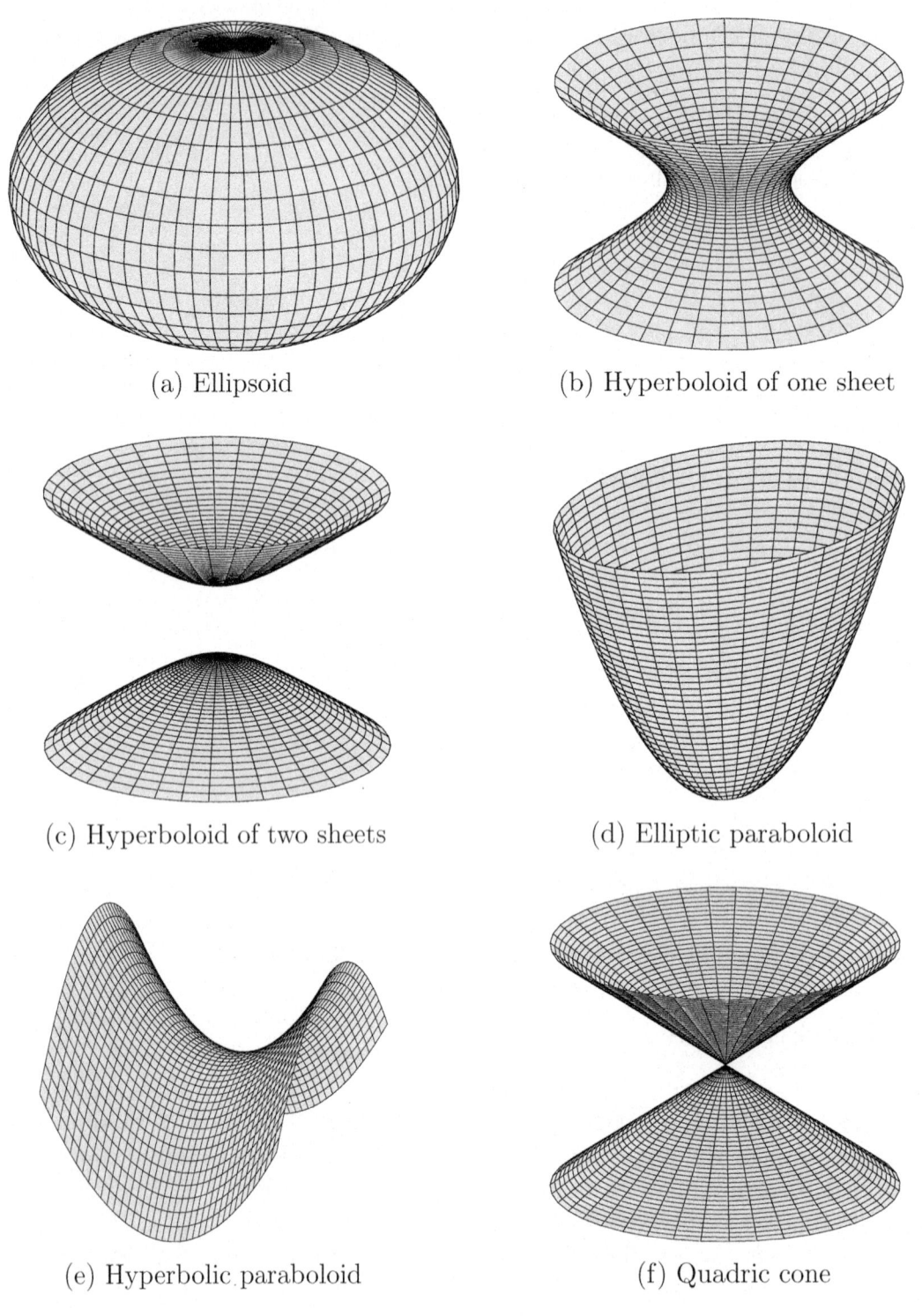

(f) Quadric cone

Figure 47: Quadratic surfaces.

## 6.4 Developable Surface

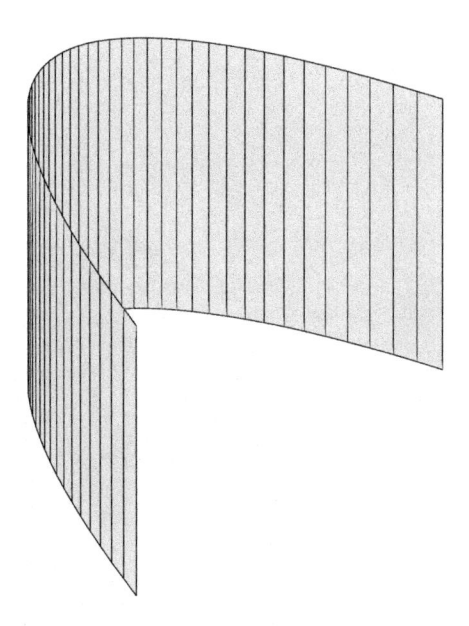

Figure 48: Parabolic cylinder as a ruled surface.

The tangent surface (see § 6.6) of a smooth curve is a ruled surface generated by the tangent line of the curve. The tangent plane is constant along a branch, represented by the tangent line at a given point, of the tangent surface of a curve. If $P$ is a point on a curve $C$ where $C$ has a tangent surface $S$, then the tangent plane to $S$ along the ruling that passes through $P$ coincides with the osculating plane of $C$ at $P$. Hence, the tangent surface may be described as the envelope of the osculating planes of the curve.

## 6.4 Developable Surface

As defined previously (see § 3.1), a surface that can be flattened into a plane without local distortion is called developable surface. A developable surface can also be defined as a surface that is isometric to the Euclidean plane. In 3D manifolds, all developable surfaces are ruled surfaces but not all ruled surfaces are developable surfaces. A ruled surface is developable if the tangent plane is constant along every ruling of the surface as it is the case with cones and cylinders. The neighborhood of each point on a sufficiently smooth surface with no flat points is developable *iff* the Gaussian curvature vanishes identically on the surface.

The generators of a developable surface and their orthogonal trajectories are its lines of curvature. A developable surface, excluding cylinder and cone, is a tangent surface of a curve where the osculating planes of the curve form the tangent planes of the surface. The collection of normal lines to a surface $S$ along a given curve $C$ on $S$ make a developable surface *iff* $C$ is a line of curvature (see § 5.8). Intrinsically, any developable surface is

---

[33] The symbols $a, b, c$ here are defined locally and hence they should not be confused with similar symbols used previously; in particular $a$ and $b$ which symbolize the determinants of the metric and curvature tensors.

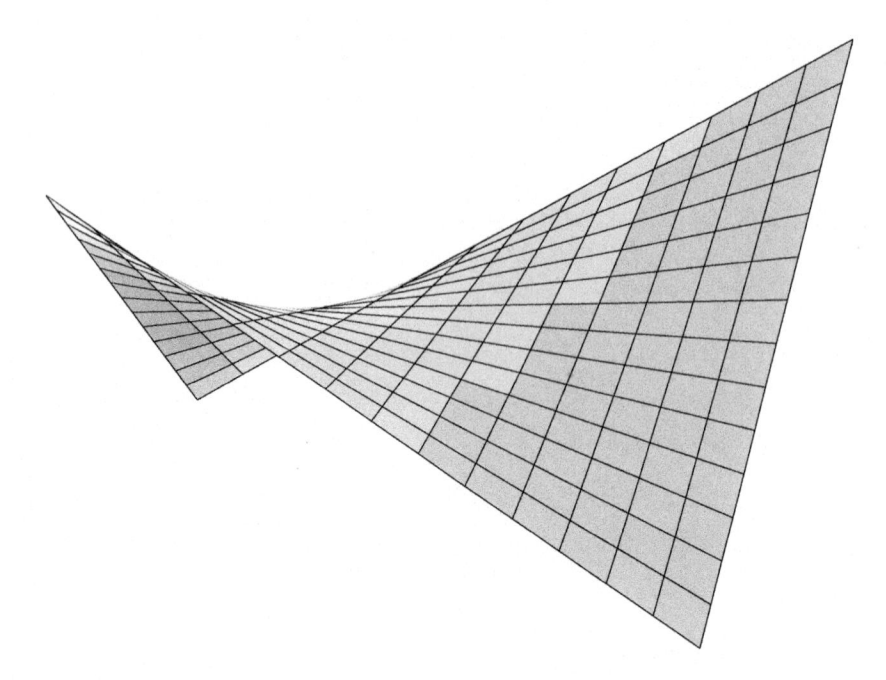

Figure 49: Hyperbolic paraboloid as a doubly-ruled surface where the grid demonstrates the two sets of straight line rulings.

equivalent (i.e. having the same metric characteristics) to a plane and hence any two developable surfaces are isometric to each other.

## 6.5 Isometric Surface

An isometry is an injective mapping from a surface $S$ to a surface $\bar{S}$ which preserves distances. If the mapping preserves distances but it is not injective it is described as local isometry. As a consequence of preserving the lengths in isometric mappings, the angles and areas are also preserved. Examples of isometric surfaces are cylinder and cone which are both isometric to plane.

Two isometric surfaces, such as a cylinder and a cone or each one of these and a plane, appear identical to a 2D inhabitant. Any difference between the two can only be perceived by an external observer residing in a reference frame in the enveloping space. Accordingly, two isometric surfaces possess identical first fundamental forms and hence any difference between them, as viewed extrinsically from the embedding space, is based on the difference between their second fundamental forms.

Isometry is an equivalence relation and hence it is reflective, symmetric and transitive, that is for three surfaces $S_1$, $S_2$ and $S_3$ we have:

1. $S_1 \sim S_1$.
2. $S_1 \sim S_2 \iff S_2 \sim S_1$.
3. If $S_1 \sim S_2$ and $S_2 \sim S_3$ then $S_1 \sim S_3$.

where the symbol $\sim$ represents an isometric relation. If two sufficiently smooth surfaces have constant equal Gaussian curvature then they are locally isometric. The mapping

6.6 Tangent Surface

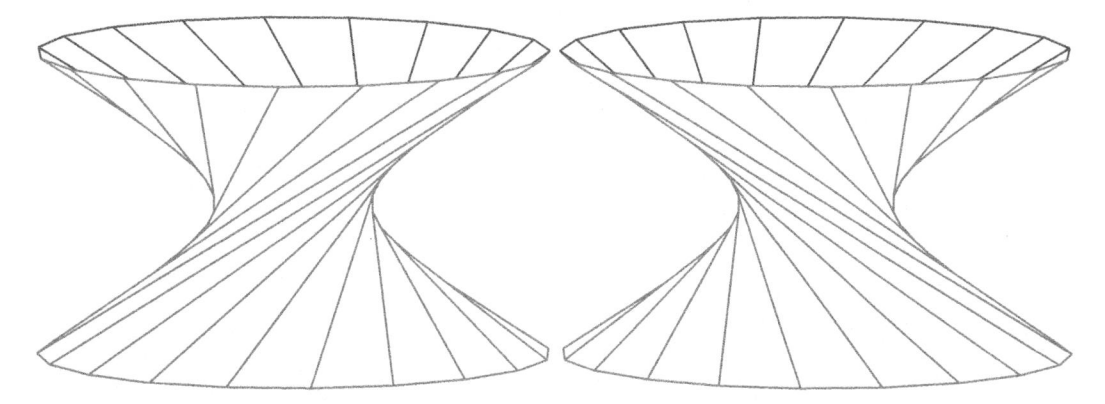

Figure 50: Hyperboloid of one sheet as a doubly-ruled surface where the two frames demonstrate the two sets of straight line rulings.

relation between the two surfaces then include three constants corresponding to the three independent coefficients of the first fundamental form. A surface of revolution is isometric to itself in infinitely many ways, each of which corresponds to a rotation of the surface through a given angle around its axis of symmetry. As indicated before, any surface is isometric to the plane *iff* the Gaussian curvature (or the Riemann-Christoffel curvature tensor) vanishes identically on the surface.

## 6.6 Tangent Surface

As stated previously, the tangent surface of a space curve is a surface generated by the assembly of all the tangent lines to the curve. The tangent lines of the curve are called the generators or branches of the tangent surface. Accordingly, the equation of a tangent surface $S_T$ to a curve $C$ is given by:

$$\mathbf{r}_T = \mathbf{r}_i + k\mathbf{T}_i \tag{440}$$

where $\mathbf{r}_T$ is an arbitrary point on the tangent surface, $\mathbf{r}_i$ is a given point on the curve $C$, $k$ is a real variable ($-\infty < k < \infty$), and $\mathbf{T}_i$ is the unit vector tangent to $C$ at $\mathbf{r}_i$. The tangent surface is generated by varying $i$ along $C$ and $k$ along the tangent line.

The tangent surface of a curve is made of two parts: one part corresponds to $k > 0$ and the other part corresponds to $k < 0$ where the curve is a border line between these two parts (see Fig. 51). The two parts of the tangent surface are tangent to each other along the curve which forms a sharp edge between the two. The curve is, therefore, called the edge of regression of the surface. The tangent plane is constant along a branch of the tangent surface of a curve. This tangent plane is the osculating plane of the curve at the point of contact of the branch with the curve. According to the definition of involute, all the involutes of a curve $C_e$ are wholly embedded in the tangent surface of $C_e$. The normal to the tangent surface of a space curve $C$ at a point of a given ruling $\mathfrak{R}$ is parallel to the binormal line of $C$ at the point of contact of $C$ with $\mathfrak{R}$.

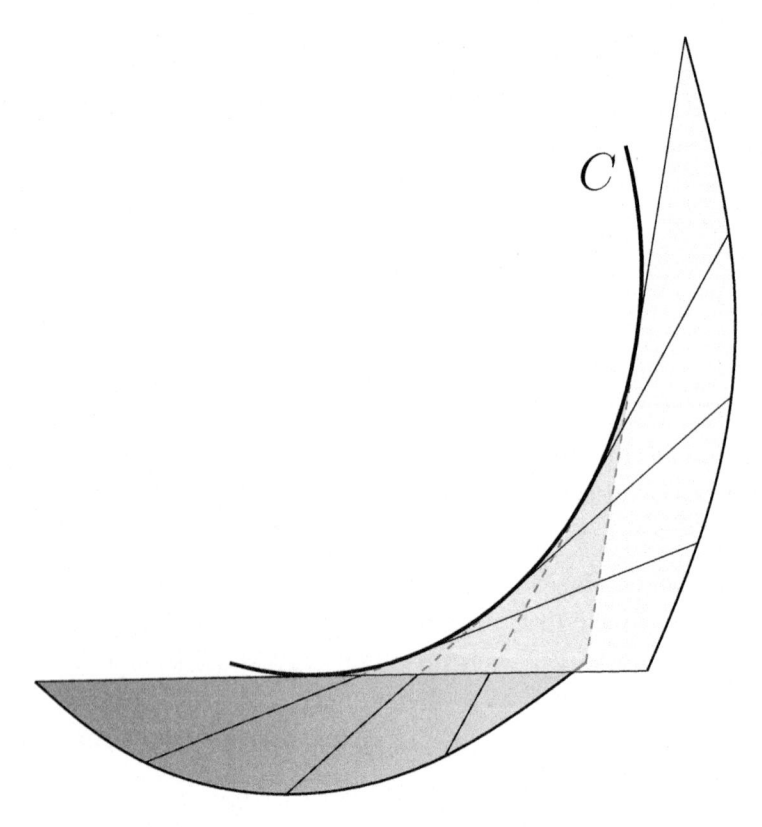

Figure 51: Space curve $C$ and the two parts of its tangent surface, shaded differently at the top and the bottom.

## 6.7 Minimal Surface

A minimal surface is a surface whose area is minimum compared to the area of any other surface sharing the same boundary. Hence, the minimal surface is an extremum with regard to the integral of area over its domain. A common physical example of a minimal surface is a soap film formed between two coaxial circular rings where it takes the minimal surface shape of a catenoid (Fig. 10) due to the surface tension. This problem, and its alike of investigations related to the physical realization of minimal surfaces, may be described as the Plateau problem. Geometric examples of minimal surface shapes are planes, catenoids, helicoids (Fig. 11) and ennepers (Fig. 13).

Since the mean curvature $H$ of a surface at a given point $P$ is a measure of the rate of change of area of the surface elements in the neighborhood of $P$, a minimal surface is characterized by having an identically vanishing mean curvature and hence the principal curvatures at each point have the same magnitude and opposite signs (see Eq. 382). A minimal surface is also characterized by having an orthogonal net of asymptotic lines and a conjugate net of minimal lines.[34] In fact, having an orthogonal net of asymptotic lines is a sufficient and necessary condition for having zero mean curvature (refer to § 5.9). Among surfaces of revolution, catenoid is the only minimal surface of this type.

---

[34] Minimal lines are curves of minimal length.

## 6.8 Exercises

6.1 State three features which are specific to plane surfaces.

6.2 Why all directions are asymptotic at any point on a plane surface?

6.3 Show that having an identically vanishing surface curvature tensor is a necessary and sufficient condition for a surface to be plane.

6.4 Show that plane is the only connected surface of class $C^2$ whose all points are flat.

6.5 Give the tensor notation form of the equation that defines quadratic surfaces.

6.6 Name three types of quadratic surface giving their canonical equation in Cartesian coordinates.

6.7 A surface is represented parametrically by: $\mathbf{r}(u,v) = (u+v, u-v, 2uv)$. Obtain the surface representation in canonical Cartesian form and hence determine its type.

6.8 Make a simple 3D plot of a hyperbolic paraboloid showing the Cartesian coordinate axes and indicating the parameters of the surface. Use a computer graphic package if convenient.

6.9 For which types of quadratic surface the origin of coordinates is not a valid point on the surface according to their canonical forms and why? Does this also apply to the non-canonical forms of these surfaces? Assuming a canonical form, are there any limiting conditions under which the origin can be included in these surfaces?

6.10 Find the parametric representation of a cylindrical surface whose intersection with the $xy$ plane is given by: $4x^2 + 9y^2 = 1$ and whose central axis is the $z$-axis. What is the type of this surface?

6.11 What is "ruled surface"? What is the other name given to this type of surface and why?

6.12 Make simple sketches for plane, cone, cylinder and Mobius strip that demonstrate their nature as ruled surfaces.

6.13 What "doubly-ruled surface" means? Give an example of such a surface with a simple sketch.

6.14 Prove that the tangent plane is constant along a branch of the tangent surface of a space curve.

6.15 Prove that the hyperbolic paraboloid is a doubly-ruled surface.

6.16 Show that a helix embedded in a circular cylinder intersects all the generators of the cylinder with constant angle.

6.17 Show that all points on the tangent surface of a given space curve are parabolic.

6.18 Justify the following statement: "At any point of a ruled surface the Gaussian curvature is non-positive". From this perspective, discuss singly- and doubly-ruled surfaces.

6.19 Prove that the tangent plane to a cylinder or a cone is constant along their generators.

6.20 Prove that if $P$ is a point on a curve $C$ where $C$ has a tangent surface $S$, then the tangent plane to $S$ along the ruling that passes through $P$ coincides with the osculating plane of $C$ at $P$.

6.21 Define "developable surface" giving several examples of this type of surface with an explanation of why they are developable surfaces.

6.22 Give an example of a ruled surface which is not developable.

## 6.8 Exercises

6.23 State the condition for a ruled surface to be developable.

6.24 Why any two developable surfaces are equivalent to each other by having the same metric characteristics?

6.25 Show that the generators of developable surfaces are lines of curvature.

6.26 Show that a necessary and sufficient condition for a ruled surface to be developable is that its Gaussian curvature vanishes identically.

6.27 What is "isometric mapping"? Give an example of such a mapping between two common types of surface.

6.28 How two isometric surfaces are seen by a 2D inhabitant? Can he distinguish between the two and why?

6.29 Prove that isometry is a symmetric relation.

6.30 Demonstrate symbolically that isometric mapping is an equivalence relation.

6.31 How two isometric surfaces are characterized in terms of their first and second fundamental forms? Provide detailed explanations.

6.32 Show that catenoid and helicoid are locally isometric.

6.33 Why a surface of revolution is isometric to itself in infinitely many ways? Demonstrate your answer by an example.

6.34 Define "tangent surface" of a space curve descriptively and mathematically.

6.35 What is the difference between the "tangent surface" of a curve and the "tangent plane" of a surface? Make detailed comparisons between the two.

6.36 Derive the equation representing the tangent surface of a space curve represented by: $\mathbf{r}(t) = (t^2, t-2, t^3+5)$.

6.37 Why the tangent surface of a space curve is made of two sections? How these sections meet on the curve?

6.38 Make a simple 3D sketch of an arbitrary twisted space curve and its tangent surface showing parts of its two sections.

6.39 Prove that the curve made by the intersection of the normal plane of a curve $C$ at a given point $P$ with the tangent surface of $C$ has a cusp at $P$.

6.40 Write down the equation representing the tangent surface of a space curve explaining all the symbols used in the equation.

6.41 In the context of tangent surface of a curve, what "branch", "generator", and "edge of regression" mean? Make an attempt to justify these names.

6.42 What is the meaning of "minimal surface"? Give geometric and physical examples of this type of surface.

6.43 Check if the surface represented parametrically by: $\mathbf{r}(\theta, \phi) = (\cosh\theta \cos\phi, \cosh\theta \sin\phi, \theta)$ is minimal or not.

6.44 Why minimal surfaces are characterized by having identically vanishing mean curvature?

6.45 What is the implication of having vanishing mean curvature $H$ on the principal curvatures of the surface at the points where $H$ vanishes?

6.46 Should a surface with orthogonal families of asymptotic lines be a minimal surface? If so, why?

# Chapter 7
# Tensor Differentiation over Surfaces

The focus of this chapter is the differentiation of tensor fields over surfaces, where general surface and space coordinates are generally assumed. In general, tensor differentiation, whether covariant or absolute, over a 2D surface follows similar rules to the rules that apply to tensor differentiation over general $n$D curved spaces. Some of these rules are:
1. The sum and product rules of differentiation apply to covariant and absolute differentiation as usual.
2. The covariant and absolute derivatives of tensors are tensors.
3. The covariant and absolute derivatives of scalars and invariant tensors of higher ranks are the same as the ordinary derivatives.
4. The covariant and absolute derivative operators commute with the contraction of indices.
5. The covariant and absolute derivatives of the metric, Kronecker and permutation tensors (and their associated tensors) vanish identically in any coordinate system, that is:

$$a_{\alpha\beta|\gamma} = 0 \qquad\qquad a^{\alpha\beta}_{\ |\gamma} = 0 \qquad (441)$$

$$\delta^{\alpha}_{\beta|\gamma} = 0 \qquad\qquad \delta^{\alpha\delta}_{\beta\omega|\gamma} = 0 \qquad (442)$$

$$\underline{\epsilon}_{\alpha\beta|\gamma} = 0 \qquad\qquad \underline{\epsilon}^{\alpha\beta}_{\ |\gamma} = 0 \qquad (443)$$

where the sign | represents covariant or absolute differentiation with respect to the surface coordinate $u^\gamma$. Hence, these tensors should be treated like constants in tensor differentiation.

An exception to these rules is the covariant derivative of the space basis vectors in their covariant and contravariant forms which is identically zero, that is:

$$\mathbf{E}_{i;j} = \partial_j \mathbf{E}_i - \Gamma^k_{ij}\mathbf{E}_k = +\Gamma^k_{ij}\mathbf{E}_k - \Gamma^k_{ij}\mathbf{E}_k = \mathbf{0} \qquad (444)$$

$$\mathbf{E}^i_{;j} = \partial_j \mathbf{E}^i + \Gamma^i_{kj}\mathbf{E}^k = -\Gamma^i_{kj}\mathbf{E}^k + \Gamma^i_{kj}\mathbf{E}^k = \mathbf{0} \qquad (445)$$

but this is not the case with the surface basis vectors in their covariant and contravariant forms, $\mathbf{E}_\alpha$ and $\mathbf{E}^\alpha$, whose covariant derivatives do not vanish identically. The reason is that, due to curvature, the partial derivatives of the surface basis vectors do not necessarily lie in the tangent plane and hence the following relations:

$$\partial_j \mathbf{E}_i = +\Gamma^k_{ij}\mathbf{E}_k \qquad (446)$$

$$\partial_j \mathbf{E}^i = -\Gamma^i_{kj}\mathbf{E}^k \qquad (447)$$

which are valid in the enveloping space and are used in Eqs. 444 and 445, are not valid on the surface anymore.

# 7 TENSOR DIFFERENTIATION

At a given point $P$ on a sufficiently smooth surface with geodesic surface coordinates and rectangular Cartesian space coordinates, the covariant and absolute derivatives reduce respectively to the partial and total derivatives at $P$.

The covariant derivative of the surface basis vectors is symmetric in its two indices, that is:

$$\begin{aligned} \mathbf{E}_{\alpha;\beta} &= \partial_\beta \mathbf{E}_\alpha - \Gamma^\gamma_{\alpha\beta} \mathbf{E}_\gamma \\ &= \partial_\alpha \mathbf{E}_\beta - \Gamma^\gamma_{\beta\alpha} \mathbf{E}_\gamma \\ &= \mathbf{E}_{\beta;\alpha} \end{aligned} \qquad (448)$$

The covariant derivative of the surface basis vectors, $\mathbf{E}_{\alpha;\beta}$, represents space vectors which are normal to the surface with no tangential component.

The covariant derivative of a differentiable rank-1 surface tensor $\mathbf{A}$ in its covariant and contravariant forms with respect to a surface coordinate $u^\beta$ is given by:

$$A_{\alpha;\beta} = \frac{\partial A_\alpha}{\partial u^\beta} - \Gamma^\gamma_{\alpha\beta} A_\gamma \qquad \text{(covariant)} \qquad (449)$$

$$A^\alpha_{;\beta} = \frac{\partial A^\alpha}{\partial u^\beta} + \Gamma^\alpha_{\gamma\beta} A^\gamma \qquad \text{(contravariant)} \qquad (450)$$

where the Christoffel symbols are derived from the surface metric.

The covariant derivative of a differentiable rank-2 surface tensor $\mathbf{A}$ in its covariant, contravariant and mixed forms with respect to a surface coordinate $u^\gamma$ is given by:

$$A_{\alpha\beta;\gamma} = \frac{\partial A_{\alpha\beta}}{\partial u^\gamma} - \Gamma^\delta_{\alpha\gamma} A_{\delta\beta} - \Gamma^\delta_{\beta\gamma} A_{\alpha\delta} \qquad \text{(covariant)} \qquad (451)$$

$$A^{\alpha\beta}_{;\gamma} = \frac{\partial A^{\alpha\beta}}{\partial u^\gamma} + \Gamma^\alpha_{\delta\gamma} A^{\delta\beta} + \Gamma^\beta_{\delta\gamma} A^{\alpha\delta} \qquad \text{(contravariant)} \qquad (452)$$

$$A^\beta_{\alpha;\gamma} = \frac{\partial A^\beta_\alpha}{\partial u^\gamma} - \Gamma^\delta_{\alpha\gamma} A^\beta_\delta + \Gamma^\beta_{\delta\gamma} A^\delta_\alpha \qquad \text{(mixed)} \qquad (453)$$

More generally, for a differentiable surface tensor $\mathbf{A}$ of type $(m, n)$, the covariant derivative with respect to a surface coordinate $u^\gamma$ is given by:

$$\begin{aligned} A^{\alpha_1 \alpha_2 \ldots \alpha_m}_{\beta_1 \beta_2 \ldots \beta_n;\gamma} &= \frac{\partial A^{\alpha_1 \alpha_2 \ldots \alpha_m}_{\beta_1 \beta_2 \ldots \beta_n}}{\partial u^\gamma} + \Gamma^{\alpha_1}_{\delta\gamma} A^{\delta \alpha_2 \ldots \alpha_m}_{\beta_1 \beta_2 \ldots \beta_n} + \Gamma^{\alpha_2}_{\delta\gamma} A^{\alpha_1 \delta \ldots \alpha_m}_{\beta_1 \beta_2 \ldots \beta_n} + \cdots + \Gamma^{\alpha_m}_{\delta\gamma} A^{\alpha_1 \alpha_2 \ldots \delta}_{\beta_1 \beta_2 \ldots \beta_n} \\ &\quad - \Gamma^\delta_{\beta_1 \gamma} A^{\alpha_1 \alpha_2 \ldots \alpha_m}_{\delta \beta_2 \ldots \beta_n} - \Gamma^\delta_{\beta_2 \gamma} A^{\alpha_1 \alpha_2 \ldots \alpha_m}_{\beta_1 \delta \ldots \beta_n} - \cdots - \Gamma^\delta_{\beta_n \gamma} A^{\alpha_1 \alpha_2 \ldots \alpha_m}_{\beta_1 \beta_2 \ldots \delta} \end{aligned} \qquad (454)$$

The covariant derivative of a space tensor with respect to a surface coordinate $u^\alpha$ is formed by the inner product of the covariant derivative of the tensor with respect to the space coordinates $x^k$ by the tensor $x^k_\alpha$. This may be considered as a form of the chain rule of differentiation. For example, the covariant derivative of $A^i$ with respect to $u^\alpha$ is given by:

$$A^i_{;\alpha} = A^i_{;k} x^k_\alpha \qquad (455)$$

The covariant derivative with respect to a surface coordinate $u^\beta$ of a mixed tensor $A^i_\alpha$, which is contravariant with respect to transformation in a space coordinate $x^i$ and covariant with respect to transformation in a surface coordinate $u^\alpha$, is given by:[35]

$$A^i_{\alpha;\beta} = \frac{\partial A^i_\alpha}{\partial u^\beta} + \Gamma^i_{jk} A^k_\alpha \frac{\partial x^j}{\partial u^\beta} - \Gamma^\gamma_{\alpha\beta} A^i_\gamma \tag{456}$$

where the Christoffel symbols with Latin and Greek indices are derived respectively from the space and surface metrics. This pattern can be easily generalized to a mixed tensor $A^{i_1...i_m}_{\alpha_1...\alpha_n}$ of type $(m,n)$ which is contravariant in transformations of space coordinates $x^i$ and covariant in transformations of surface coordinates $u^\alpha$. For example, the covariant derivative of a tensor $A^{ij}_{\alpha\beta}$ with respect to $u^\gamma$ is given by:

$$A^{ij}_{\alpha\beta;\gamma} = \frac{\partial A^{ij}_{\alpha\beta}}{\partial u^\gamma} + \Gamma^i_{mk} A^{mj}_{\alpha\beta} \frac{\partial x^k}{\partial u^\gamma} + \Gamma^j_{mk} A^{im}_{\alpha\beta} \frac{\partial x^k}{\partial u^\gamma} - \Gamma^\delta_{\alpha\gamma} A^{ij}_{\delta\beta} - \Gamma^\delta_{\beta\gamma} A^{ij}_{\alpha\delta} \tag{457}$$

The above rules can be extended further to include tensors with space and surface contravariant indices and space and surface covariant indices. For Example, the covariant derivative of a tensor $A^{i\alpha}_{j\beta}$ with respect to a surface coordinate $u^\gamma$, where $i$ and $j$ are space indices and $\alpha$ and $\beta$ are surface indices, is given by:

$$A^{i\alpha}_{j\beta;\gamma} = \frac{\partial A^{i\alpha}_{j\beta}}{\partial u^\gamma} + \Gamma^i_{mk} A^{m\alpha}_{j\beta} \frac{\partial x^k}{\partial u^\gamma} + \Gamma^\alpha_{\delta\gamma} A^{i\delta}_{j\beta} - \Gamma^m_{jk} A^{i\alpha}_{m\beta} \frac{\partial x^k}{\partial u^\gamma} - \Gamma^\delta_{\beta\gamma} A^{i\alpha}_{j\delta} \tag{458}$$

This example can be easily extended to the most general form of a tensor with any combination of covariant and contravariant space and surface indices.

The covariant derivative of the surface basis vector $\mathbf{E}_\alpha$, which in tensor notation is denoted by $x^i_\alpha$, is given by:

$$x^i_{\alpha;\beta} = \frac{\partial^2 x^i}{\partial u^\beta \partial u^\alpha} + \Gamma^i_{jk} x^j_\alpha x^k_\beta - \Gamma^\delta_{\alpha\beta} x^i_\delta \tag{459}$$

From the last equation, we conclude:

$$x^i_{\alpha;\beta} = x^i_{\beta;\alpha} \tag{460}$$

This is because the Christoffel symbols are symmetric in their paired indices (see Eq. 66) and we have $\partial_{\alpha\beta} x^i = \partial_{\beta\alpha} x^i$ and $x^j_\alpha x^k_\beta = x^k_\beta x^j_\alpha$. This symmetry has also been established earlier using vector notation for the surface basis vectors (see Eq. 448).

The mixed second order covariant derivative of the surface basis vectors is given by:

$$x^i_{\alpha;\beta\gamma} = b_{\alpha\beta;\gamma} n^i + b_{\alpha\beta} n^i_{;\gamma} = b_{\alpha\beta;\gamma} n^i - b_{\alpha\beta} a^{\delta\omega} b_{\delta\gamma} x^i_\omega \tag{461}$$

where the covariant derivative of the surface covariant curvature tensor is given, as usual, by:

$$b_{\alpha\beta;\gamma} = \frac{\partial b_{\alpha\beta}}{\partial u^\gamma} - \Gamma^\delta_{\alpha\gamma} b_{\delta\beta} - \Gamma^\delta_{\beta\gamma} b_{\alpha\delta} \tag{462}$$

---

[35] An example of such a tensor is $x^i_\alpha$ which was discussed earlier, e.g. in § 3.3.

# 7 TENSOR DIFFERENTIATION

The covariant differentiation operators in mixed derivatives are not commutative and hence for a contravariant surface vector $A^\gamma$, for instance, we have:

$$A^\gamma_{;\alpha\beta} - A^\gamma_{;\beta\alpha} = R^\gamma_{\delta\alpha\beta} A^\delta \tag{463}$$

where $R^\gamma_{\delta\alpha\beta}$ is the Riemann-Christoffel curvature tensor of the second kind for the surface. Similarly, the mixed second order covariant derivatives of the surface basis vectors satisfy the following relation:

$$x^i_{\alpha;\beta\gamma} - x^i_{\alpha;\gamma\beta} = R^\delta_{\alpha\beta\gamma} x^i_\delta \tag{464}$$

This, in fact, is an instance of the general relation: $A_{j;kl} - A_{j;lk} = R^i_{jkl} A_i$ which is found in tensor calculus texts.

As defined in tensor calculus books, the absolute or intrinsic derivative of a tensor field along a $t$-parameterized curve in an $n$D space with respect to the parameter $t$ is the inner product of the covariant derivative of the tensor and the tangent vector to the curve. This identically applies to the absolute derivative of curves contained in 2D surfaces.

Consequently, the absolute derivative of a tensor field along a $t$-parameterized curve on a surface with respect to the parameter $t$ follows similar rules to those of a space curve in a general $n$D space, as stated in the literature of tensor calculus. For example, the absolute derivative of a differentiable surface vector field **A** in its covariant and contravariant forms with respect to the parameter $t$ is given by:

$$\frac{\delta A_\alpha}{\delta t} = \frac{dA_\alpha}{dt} - \Gamma^\gamma_{\alpha\beta} A_\gamma \frac{du^\beta}{dt} \tag{465}$$

$$\frac{\delta A^\alpha}{\delta t} = \frac{dA^\alpha}{dt} + \Gamma^\alpha_{\gamma\beta} A^\gamma \frac{du^\beta}{dt} \tag{466}$$

where the Christoffel symbols are derived from the surface metric. It should be remarked that if **A** is a space vector field defined along the above surface curve then the above formulae will take a similar form but with change from surface to space coordinates, and hence the curve is treated as a space curve, that is:

$$\frac{\delta A_i}{\delta t} = \frac{dA_i}{dt} - \Gamma^j_{ik} A_j \frac{dx^k}{dt} \tag{467}$$

$$\frac{\delta A^i}{\delta t} = \frac{dA^i}{dt} + \Gamma^i_{jk} A^j \frac{dx^k}{dt} \tag{468}$$

where the Christoffel symbols are derived from the space metric.

The absolute derivative of the tensor $A^i_\alpha$, which is defined in the previous paragraphs, along a $t$-parameterized surface curve is given by:

$$\frac{\delta A^i_\alpha}{\delta t} = A^i_{\alpha;\beta} \frac{du^\beta}{dt} = \left( \frac{\partial A^i_\alpha}{\partial u^\beta} + \Gamma^i_{jk} A^k_\alpha \frac{\partial x^j}{\partial u^\beta} - \Gamma^\gamma_{\alpha\beta} A^i_\gamma \right) \frac{du^\beta}{dt} \tag{469}$$

The pattern of absolute differentiation, as seen in the above examples, can be easily extended to a more general type of tensor with covariant and contravariant space and surface indices, as done for covariant differentiation.

To extend the idea of geodesic coordinates to deal with mixed tensors of the type $A^i_\alpha$, a rectangular Cartesian coordinate system over the space and a geodesic system on the surface are introduced and hence at the poles the absolute and covariant derivatives become total and partial derivatives respectively.

As indicated above, the covariant and absolute derivatives of space and surface metric, permutation and Kronecker tensors in their covariant, contravariant and mixed forms vanish identically and hence they behave as constants with respect to tensor differentiation when involved in inner or outer product operations with other tensors and commute with these operators. Similarly, the *surface* covariant and absolute derivatives of *space* metric tensor, space Kronecker tensor, space permutation tensor and space basis vectors vanish identically, that is:

$$g_{ij|\gamma} = 0 \qquad\qquad g^{ij}_{\phantom{ij}|\gamma} = 0 \qquad (470)$$

$$\delta^i_{j|\gamma} = 0 \qquad\qquad \delta^{ij}_{kl|\gamma} = 0 \qquad (471)$$

$$\underline{\epsilon}_{ijk|\gamma} = 0 \qquad\qquad \underline{\epsilon}^{ijk}_{\phantom{ijk}|\gamma} = 0 \qquad (472)$$

$$\mathbf{E}_{i|\gamma} = \mathbf{0} \qquad\qquad \mathbf{E}^i_{\phantom{i}|\gamma} = \mathbf{0} \qquad (473)$$

where the sign | represents covariant or absolute differentiation with respect to the surface coordinate $u^\gamma$. Hence, these space tensors are in lieu of constants with respect to surface tensor differentiation.

The nabla $\nabla$ based differential operations, such as gradient and divergence, apply to space surface as for any general curved space and hence the formulae given in the literature of tensor calculus for a general $n$D space can be used with the substitution of the surface coordinates and surface metric parameters. For example, the divergence of a surface vector field $A^\alpha$ is given by:

$$\nabla \cdot \mathbf{A} = \frac{1}{\sqrt{a}} \partial_\alpha \left( \sqrt{a} A^\alpha \right) \qquad (474)$$

and the Laplacian of a surface scalar field $f$ is given by:

$$\nabla^2 f = \frac{1}{\sqrt{a}} \partial_\alpha \left( \sqrt{a} a^{\alpha\beta} \partial_\beta f \right) \qquad (475)$$

where $a^{\alpha\beta}$ is the surface contravariant metric tensor, $a$ is the determinant of the surface covariant metric tensor and $\alpha, \beta = 1, 2$.

## 7.1 Exercises

7.1 Summarize the main rules that govern the differentiation of tensor fields over surfaces and compare these rules to those of $n$D spaces ($n > 2$).

7.2 Is there any rule of tensor differentiation that applies to $n$D spaces ($n > 2$) but not to surfaces? If so, which and why? State your answer with a full formal explanation.

7.3 At the points of a smooth surface with geodesic surface coordinates and Cartesian spatial coordinates of a flat embedding 3D space, what happens to the covariant and absolute derivatives of tensor fields?

## 7.1 Exercises

7.4 Derive the following identity: $\mathbf{E}_{\alpha;\beta} = \mathbf{E}_{\beta;\alpha}$.

7.5 Express the identity in the previous exercise in full tensor notation.

7.6 Is $\mathbf{E}_{\alpha;\beta}$ a surface vector or a space vector? Discuss the possible different meanings of these attributes.

7.7 Explain, in detail, the following equation related to the covariant derivative of space tensors with respect to surface coordinates: $A^i_{;\alpha} = A^i_{;k} x^k_\alpha$.

7.8 Write down the mathematical expression for the covariant derivative of the tensor $A^{\delta j}_{kn\beta}$ with respect to the surface coordinate $u^\gamma$ where the Latin and Greek indices represent space and surface general coordinates.

7.9 Write down the tensor equation for the covariant derivative of the surface basis vector $x^m_\gamma$ with respect to the index $\beta$.

7.10 Complete the following equation which involves the tensor $\mathbf{B}$ where the indices represent surface coordinates:
$$B^\alpha_{;\gamma\delta} - B^\alpha_{;\delta\gamma} = ?$$

7.11 Give a brief descriptive definition of the absolute differentiation of a tensor field along a curve.

7.12 What is the other name given to the absolute differentiation?

7.13 Explain the mathematical pattern of absolute differentiation of tensor fields along surface curves illustrating this by an example.

7.14 Is the pattern of absolute differentiation of surface tensor fields along surface curves identical to the pattern of absolute differentiation of space tensor fields along space curves? If there is any difference, identify and explain.

7.15 Write, in expanded form, the mathematical equation of the following intrinsic derivative: $\frac{\delta B^k_\gamma}{\delta t}$ where $\mathbf{B}$ is a tensor and $k$ and $\gamma$ are space and surface indices.

7.16 What are the covariant and absolute derivatives of space and surface metric, permutation and Kronecker tensors in their covariant, contravariant and mixed forms?

7.17 Do the operators of covariant and absolute differentiation with respect to space and surface coordinates commute with the metric tensor involved in an inner or outer product with another tensor? Explain why.

7.18 Explain the following identity giving detailed definitions of all the symbols and notations involved: $\epsilon_{ijk|\gamma} = 0$.

7.19 Do the nabla based differential operations apply to the surface tensor fields as to the tensor fields in curved spaces of higher dimensionality?

7.20 What is the Laplacian of a differentiable coordinate-dependent surface scalar field $h$ (i.e $\nabla^2 h$)? Write in your answer the mathematical equation for this operation defining all the symbols used.

7.21 Compare the equation in the previous question with the equation of the Laplacian of a scalar field defined over a general $n$D space.

# References

L.P. Eisenhart. *An Introduction to Differential Geometry.* Princeton University Press, second edition, 1947.

J. Gallier. *Geometric Methods and Applications for Computer Science and Engineering.* Springer, second edition, 2011.

P. Grinfeld. *Introduction to Tensor Analysis and the Calculus of Moving Surfaces.* Springer, first edition, 2013.

J.H. Heinbockel. *Introduction to Tensor Calculus and Continuum Mechanics.* 1996.

D.C. Kay. *Schaum's Outline of Theory and Problems of Tensor Calculus.* McGraw-Hill, first edition, 1988.

R. Koch. *Mathematics 433/533; Class Notes.* University of Oregon, 2005.

E. Kreyszig. *Differential Geometry.* Dover Publications, Inc., second edition, 1991.

M.M. Lipschutz. *Schaum's Outline of Differential Geometry.* McGraw-Hill, first edition, 1969.

T. Sochi. *Tensor Calculus Made Simple.* CreateSpace, first edition, 2016.

I.S. Sokolnikoff. *Tensor Analysis Theory and Applications.* John Wiley & Sons, Inc., first edition, 1951.

B. Spain. *Tensor Calculus: A Concise Course.* Dover Publications, third edition, 2003.

J.L. Synge; A. Schild. *Tensor Calculus.* Dover Publications, 1978.

# Index

1D inhabitant, 12, 52, 62
2D inhabitant, 11, 12, 39, 86, 88, 89, 107, 146, 156, 159, 169, 176, 180
3D inhabitant, 146

Absolute
    derivative, 5, 33, 47, 48, 57, 59, 64, 159, 181, 182, 184–186
    differentiation, 181, 184–186
    permutation tensor, 7, 38, 48, 49, 74, 83, 84, 129, 130, 162
    tensor, 36, 76, 83, 129
    value, 55, 101, 125, 127, 165
Affine connection, 34
Algebra, 1
Algebraic geometry, 8
Ambient space, 11, 43, 104, 107, 110, 146
Analytic curve, 33
Angle, 7, 12, 13, 55, 59, 60, 69, 70, 78, 82, 83, 88, 104, 106, 112, 115, 117, 118, 120, 127, 133–136, 148, 152, 160, 161, 163, 165, 168–170, 176, 177, 179
Anti-
    parallel, 44, 53, 71, 110, 160
    symmetric, 37, 38, 42, 56
Antipodal point, 157
Apex, 28, 126, 137
Arc length, 6, 9, 20, 43, 44, 46, 48, 51–57, 60, 78–80, 87, 88, 104, 106, 155, 157, 159
Area, 5, 7, 12, 20, 27, 70, 78, 80, 81, 83, 88, 101, 102, 104, 106, 120, 127, 128, 131–133, 136, 148, 176, 178
    integral, 120, 128, 133, 136, 148, 178
Asymptote, 18, 140
Asymptotic
    curve, 164
    direction, 115, 163–166, 170, 172, 179
    line, 55, 150, 163–166, 170, 171, 178, 180
    polygonal arc, 20, 27
    polygonal surface, 20, 27
Axis
    of coordinates, 13, 16, 18, 24, 25, 40, 41, 125, 146, 168, 173, 179
    of revolution, 12, 13
    of symmetry, 12, 39, 94, 138, 173, 177, 179

Basis vectors, 5, 7, 9, 28–30, 33, 36, 41, 42, 47, 49, 50, 56, 61, 66, 71–76, 78, 80, 90, 97–99, 102–105, 107, 108, 118, 130, 144, 181–186
Beltrami
    -Enneper theorem, 164, 170
    pseudo-sphere, 18, 126, 146
Bending, 68
Bertrand curve, 150–152, 167, 168
Bicontinuous, 20, 40, 65, 68, 69
Binormal
    indicatrix, 168
    line, 49, 61, 63, 150, 167, 177
    vector, 5, 7, 30, 47–49, 51, 59, 61, 97, 153
Bonnet
    formula, 55
    theorem, 93
Branch, 71, 104, 175, 177, 179, 180

Calculus, 1, 8, 12, 85
    of variations, 168
Cartesian coordinate system, 7–9, 13, 19, 25, 30, 32, 33, 40, 47, 48, 52, 54, 56, 60, 62, 65, 71, 76, 84, 87, 98, 104, 105, 110, 121, 173, 179, 182, 185
Catenary, 16, 40, 62, 106
Catenoid, 16, 20, 39, 138, 146, 178, 180
Center
    of curvature, 57, 58, 63, 115, 116, 119, 120, 145, 150, 151, 154, 167
    of spherical curvature, 59
Chain rule of differentiation, 182
Christoffel symbol, 7, 9, 32–38, 42, 78, 85, 86, 92, 100, 105, 116, 127, 159, 182–184
Circle, 12, 13, 22, 39, 45, 46, 52, 57, 58, 60–62, 94, 128, 135, 138, 140, 151, 153, 154, 156, 158, 160, 167, 168
    of curvature, 57
Cissoid of Diocles, 60
Closed
    curve, 46, 60, 67, 160, 168
    surface, 6, 19, 60, 68, 102, 103, 137, 146, 158
Codazzi
    -Mainardi equations, 92, 100, 101, 109
    equation, 100, 101
Collinear, 55, 110, 114, 120, 155, 161, 164
Comma notation, 5
Commutative, 184
Commute, 181, 185, 186
Compact surface, 19, 40, 68, 69, 102, 103, 126, 136, 137, 145, 147

188

Compression, 67, 69, 125
Cone, 28, 40, 68, 69, 102, 103, 106, 126, 128, 137, 146, 168, 173, 175, 176, 179
Conformal mapping, 69, 70, 83, 102–104
Conjugate
    direction, 165, 166, 170, 171
    grid, 178
    hyperbola, 94, 140
    roots, 165
Connected surface, 41, 68, 102, 126, 145–147, 172, 179
Continuous, 20, 21, 25, 39, 45, 60, 68, 71, 87, 90, 105, 123, 133, 153, 173
Contraction of indices, 181
Contravariant, 5, 7, 29, 30, 34, 36, 38, 49, 74–78, 83–85, 94, 98, 104–106, 109, 181–186
Coordinate
    curve, 7, 26, 28, 30, 33, 35, 41, 66, 67, 72–75, 79, 81, 83, 104, 114, 117, 118, 123, 127, 144, 145, 159–164, 166, 169–171
    grid, 72, 104
    patch, 33, 65, 67, 68, 102, 127, 163
Corner of curve, 133, 134, 136, 160
Cosine of angle, 82, 112
Covariant, 5, 7, 10, 29, 30, 34–36, 38, 39, 48, 69, 72–80, 82–88, 90–94, 97–101, 104–109, 113, 114, 116, 123, 124, 127, 129, 138, 144, 162, 165, 170, 181–186
    derivative, 5, 33, 59, 98, 100, 101, 181–186
    differentiation, 181, 184–186
Cross product, 73, 75, 111, 112
Cube, 25
Curvature
    direction, 119
    invariant, 38
    of curve, 7, 48, 51–54, 56, 57, 59–62, 97, 111, 112, 114–118, 150–156
    scalar, 38
    tensor, 5, 10, 11, 39, 52, 62, 83–86, 91, 92, 94, 97–101, 106–109, 113, 114, 123, 124, 129–132, 138, 141, 144, 146, 147, 149, 162, 165, 170, 172, 179, 183
    vector, 7, 51, 55, 110, 111, 113, 114, 116, 142–144, 160, 164
Curved space, 30–32, 37, 40, 41, 158, 161, 181, 185, 186
Curvilinear
    coordinate system, 25, 63, 65, 105
    coordinates, 9, 33, 37, 47, 48, 54, 57
    polygon, 19, 26, 136
Cylinder, 40, 68, 69, 88, 93, 102, 103, 105, 120, 121, 125, 126, 128, 132, 137, 138, 146–148, 156, 158, 160, 168, 170, 173, 175, 176, 179
Cylindrical coordinates, 7, 13

Darboux
    frame, 5, 55, 123, 145
    vector, 5, 56, 63
Degenerate, 13, 114, 156, 158, 172
Determinant, 5, 23, 78, 85, 122, 124
    of curvature tensor, 5, 84–86, 99, 106, 107, 113, 124, 138, 141, 144, 149, 165, 170
    of metric tensor, 5, 35, 36, 38, 73, 74, 78, 80, 84, 85, 90, 93, 98, 105, 107, 116, 124, 127, 141, 149, 185
Developable surface, 69, 103, 125, 128, 147, 161, 162, 169, 175, 176, 179, 180
Diagonal, 30, 32, 65, 73
Differentiable, 21, 24, 43, 45, 51, 67–69, 71, 81, 103, 146, 182, 184, 186
Differential
    calculus, 8
    equation, 8, 18, 57, 92, 158, 159, 164, 169
    geometry, 1, 8–12, 27, 30–32, 36, 39, 41, 51, 62, 76, 92, 93, 102, 108, 112, 123, 126, 131, 158, 172
    operation, 185, 186
    operator, 5
    topology, 8
Direct conformal mapping, 69, 103
Disc, 22, 133, 134, 148
Discriminant, 7, 122, 142, 165
Distance, 11, 13, 20, 29, 51, 69, 70, 78, 97, 110, 151, 152, 155–157, 168, 176
Distortion, 46, 69, 125, 150, 175
Divergence operation, 185
Dodecahedron, 40
Dot product, 35, 53, 54, 60, 73, 74, 83, 88, 91, 107, 111, 126, 166
Double-side surface, 10
Doubly-ruled surface, 173, 176, 177, 179
Dupin indicatrix, 93–96, 108, 140, 149

Edge
    of polyhedron, 5, 19, 40, 136
    of regression, 177, 180
Element of
    arc, 79, 80, 87, 106, 157
    area, 128
    curve, 5
    surface, 5, 80, 81, 106, 131, 178
Elementary surface, 68, 103, 172
Ellipse, 39, 45, 46, 60, 94, 140
Ellipsoid, 13, 15, 19, 20, 31, 32, 40, 68, 103, 120, 121, 137, 138, 141, 146, 148, 172–174

Elliptic
- paraboloid, 14, 18, 32, 40, 41, 68, 69, 103, 114, 140, 141, 172–174
- point, 94–96, 108, 114, 137–140, 144, 148, 149, 163, 165, 166, 171
- umbilic point, 141

Embedding space, 11, 31, 43, 47, 62, 88, 97, 123, 176
Enneper surface, 17, 23, 40, 178
Enveloping space, 30, 43, 76, 87, 88, 105, 176, 181
Euclidean
- motion, 51
- plane, 37, 175
- space, 6, 20, 27, 30–32, 37, 41, 47, 52, 59, 64, 65, 76, 104, 105, 150, 155, 158, 159, 173

Euler
- -Lagrange variational principle, 23, 155
- -Poincare characteristic, 19
- characteristic, 7, 19, 20, 25, 40, 133, 136, 137, 148
- equation, 165
- theorem, 120, 145

Evolute, 5, 71, 104, 150–152, 167
Exterior angle, 133, 136
Extremum, 118, 155, 178
Extrinsic
- geometry, 32, 89, 92, 108, 146
- property, 10–12, 39, 52, 86, 104, 106, 112, 114, 116, 126, 143, 144

Face of polyhedron, 5, 19, 40, 136
Finite, 19, 68, 133
First
- curvature, 51
- derivative, 6, 63
- fundamental form, 5, 6, 10–12, 35, 37, 39, 52, 62, 69, 70, 73, 77, 78, 80, 85–93, 96–99, 101, 105–109, 112–115, 118, 121–125, 127, 128, 130–132, 138, 141, 143, 144, 147, 159, 161, 163, 165, 176, 177, 180
- fundamental quadratic form, 87
- groundform, 76
- order differential equation, 57
- variation, 23, 155

Flat
- point, 94, 102, 108, 114, 137, 138, 140, 141, 143, 148, 149, 163, 172, 179
- space, 8, 30–33, 36, 37, 40, 41, 47, 59, 76, 86, 87, 110, 158, 161, 185
- surface, 22, 37
- umbilical point, 141, 172

Frenet
- -Serret formulae, 49, 52–54, 56, 57, 62, 63, 97, 108
- formulae, 56
- frame, 49, 50, 61
- trihedron, 49

Functional mapping, 20, 25, 27, 45, 65
Fundamental
- surface tensor, 76
- theorem of curves, 51, 52, 57, 62, 92, 93
- theorem of surfaces, 52, 92, 93, 108

Gauss
- -Bonnet theorem, 102, 120, 127, 133–137, 148, 160, 169
- -Codazzi equation, 101
- equations, 97, 99, 100, 108, 109
- mapping, 101, 168

Gaussian
- coordinates, 9
- curvature, 6, 11, 31, 32, 37, 38, 42, 60, 69, 85, 96, 98, 99, 101, 102, 107, 109, 120–122, 124–133, 135–138, 142, 145–149, 160, 163–165, 169, 170, 172, 173, 175–177, 179, 180

General
- coordinates, 9, 76, 87, 186
- parameter, 5, 6, 9, 21, 44, 45, 53, 54, 60, 61, 63, 117, 150

Generalized Kronecker delta, 7
Generator, 71, 102, 103, 151, 158, 162, 170, 175, 177, 179, 180
Genus, 6, 20, 26, 40, 133, 137, 148
Geodesic
- component, 7, 55, 111, 114, 116, 143, 144, 160
- coordinates, 32–34, 127, 147, 185
- curvature, 7, 55, 111, 112, 116–118, 132, 133, 136, 143, 144, 154, 156, 160, 161
- curvature vector, 116, 159
- curve, 33, 55, 127, 132, 136, 148, 150, 154–162, 167–169
- normal vector, 6, 112, 118
- polygon, 136, 148
- system, 32–34, 185
- torsion, 7, 55, 56, 63, 162, 170
- triangle, 135, 136, 160

Geometric
- invariant, 136
- property, 138

Geometry, 1, 8, 10, 12, 29, 31, 36, 44, 88, 133
Global property, 10, 11, 39, 133
Gradient operation, 75, 185
Great circle, 115, 134, 155–158, 160

Handle, 20, 137

Helicoid, 17, 21, 40, 146, 178, 180
Helix, 13, 14, 17, 40, 52, 61, 62, 151–153, 155, 156, 158, 160, 167, 168, 179
Hole, 20, 67
Homeomorphic, 137
Homogeneous
    coordinate system, 32, 41
    linear equations, 122
Hyperbolic
    cosine, 16
    paraboloid, 15, 18, 19, 32, 39, 68, 114, 128, 140, 146, 147, 172–174, 176, 179
    point, 94–96, 108, 114, 137–141, 144, 148, 149, 163, 165, 166, 171
Hyperboloid
    of one sheet, 14, 16, 39, 103, 120, 121, 145, 172–174, 177
    of two sheets, 14, 17, 39, 103, 172–174

Icosahedron, 40
Identity
    tensor, 77
    transformation, 25
Index
    lowering, 34, 77
    lowering operator, 77
    raising, 34, 38, 77
    raising operator, 36, 38, 77, 85
    shifting operator, 37, 75, 78
Infinite, 25, 33, 111, 141, 151, 157, 167
Infinitesimal, 5, 25, 72, 79–81, 106, 128, 157
Inflection point, 49, 111, 142, 164
Injective, 43, 45, 61, 65, 69, 70, 81, 102, 109, 176
Inner product, 182, 184–186
Integral calculus, 8
Integration, 57, 63
Interior
    angle, 134–136, 148, 160
    point, 45, 71, 81
Intrinsic
    derivative, 184, 186
    distance, 29, 41
    equations, 51, 62
    geometry, 12, 32, 36, 91, 108, 146
    property, 10–12, 29, 36, 37, 39, 52, 70, 80, 86, 88, 89, 101, 104, 106, 112, 116, 117, 124–126, 132, 136, 143, 144, 146, 159, 161
Invariance, 33, 138, 161
Invariant, 24, 29, 41, 51, 55, 69, 70, 80, 85, 91, 114, 119, 120, 124–126, 131, 136, 138, 148, 149, 160, 181
Inverse
    conformal mapping, 69, 103

    function, 21
    mapping, 21, 51, 69, 71, 103
    of matrix, 77
Involute, 5, 71, 104, 151, 152, 167, 177
Isometric, 46
    mapping, 29, 41, 51, 69, 70, 103, 104, 146, 176, 180
    surface, 69, 70, 83, 107, 126, 161, 176, 177, 180
    transformation, 46, 80, 91, 126, 161
Isometry, 69, 104, 176, 180

Jacobian, 6, 23, 24, 40, 78, 84, 138
    matrix, 6, 23, 27, 28

Kissing circle, 57
Klein bottle, 68, 103
Kronecker delta, 7, 34, 75, 105, 181, 185, 186

Lancret
    equation, 51, 61
    theorem, 61
Laplacian operator, 5, 185, 186
Length, 5, 6, 12, 18, 20, 27, 29, 30, 41, 44, 47, 51, 61, 69, 70, 74, 79, 80, 83, 87, 88, 104, 106, 110, 132, 135, 151, 155–158, 176, 178
Limit, 20, 45, 50, 58, 63, 101, 157
Line
    element, 5, 30, 32, 51, 79
    of curvature, 56, 150, 156, 161–163, 169, 170, 175, 180
Linear
    algebra, 28, 85
    combination, 48, 50, 66, 72, 76, 98
    equations, 122
    transformation, 24, 32
Linearly
    dependent vectors, 150, 167
    independent vectors, 30, 66, 71, 74, 104, 123
Local
    isometry, 29, 41, 70, 104, 126, 176, 180
    property, 10, 11, 39
    shape of surface, 93, 94, 137, 138, 140, 148, 149
Lorentz transformations, 32

Mapping, 20, 24, 28, 29, 43, 65, 69–71, 83, 101–104, 109, 153, 168, 176, 180
Matrix notation, 91, 131
Mean curvature, 6, 85, 96, 99, 102, 107, 109, 120–122, 127, 131, 132, 142, 145, 147–149, 165, 170, 172, 178, 180
Meridians, 12, 13, 39, 134, 156, 162, 169
    of longitude, 13, 157
Metric tensor, 5, 6, 9–11, 30, 32, 34–39, 42, 46, 48, 60, 69, 73–82, 84–88, 90, 92–94, 97, 98,

100, 105–109, 113, 114, 116, 123, 124, 126, 127, 129, 141, 144, 146, 149, 159, 162, 176, 180–186

Meusnier theorem, 112, 115, 116, 143, 144
Minimal surface, 102, 132, 163, 172, 178, 180
Minkowski space, 30, 32, 88
Mixed
    derivative, 183, 184
    tensor, 7, 29, 34, 75–78, 84, 85, 98, 105, 107, 123, 124, 131, 147, 182, 183, 185, 186
Mobius strip, 10, 11, 68, 103, 173, 179
Monge patch, 67, 78, 80, 81, 85, 88, 91, 99, 100, 102, 103, 105–109, 125, 131, 132, 146, 147, 159, 169, 170
Monkey saddle, 17, 22, 39, 128
Moving
    frame, 112, 123, 143
    trihedron, 49
Mutually
    orthogonal, 25, 30, 41, 61, 97
    perpendicular, 39, 47, 49, 50

nabla operator, 5, 185, 186
Natural
    equations, 51, 62
    parameter, 5, 6, 9, 44, 52, 55, 57, 59–62, 127, 151, 153, 154, 159
    parameterization, 44, 52, 60, 110
Navel point, 141
Negative
    curvature, 32, 51, 102, 114, 118, 126, 128, 129, 135, 136, 140, 146, 148, 160, 163, 169
    orthogonal transformation, 24, 25, 40
Non-
    asymptotic direction, 115
    Cartesian, 19
    degenerate, 172
    Euclidean space, 64
    geodesic coordinates, 33
    linear differential equation, 159
    negative, 51, 53, 62, 122
    orientable surface, 68, 103
    periodic, 46
    planar curve, 61, 136
    polyhedral surface, 19, 40
    positive, 126, 127, 145, 163, 173, 179
    regular point, 49
    straight line, 164
    umbilical point, 55, 114, 119, 120, 122, 123, 162, 163, 165, 169
Normal
    component, 7, 98, 101, 111, 113, 114, 116, 142, 143, 164

    curvature, 7, 91, 111–116, 118–120, 125, 140, 141, 143, 145, 163, 164, 166
    indicatrix, 168
    line to surface, 67, 103, 175
    plane, 50, 51, 61, 112, 154, 180
    section, 110, 112, 114–116, 118–121, 125, 141, 145
    vector to surface, 6, 30, 36, 39, 41, 55, 59, 66–68, 72–75, 86, 89, 91, 93, 94, 97–99, 101, 103–105, 107, 108, 110, 112, 114, 115, 117, 118, 120, 121, 123, 125, 126, 131, 132, 138, 143, 146, 147, 149, 155, 160, 164, 169

Octahedron, 25
One-side surface, 10
One-to-one, 28, 65, 69, 71, 101
Ordinary
    derivative, 47, 181
    differential equation, 92
Orientable surface, 19, 20, 40, 68, 103, 110, 136, 137
Oriented
    curve, 43, 69
    surface, 68, 103, 123
Origin of coordinates, 39, 41, 101, 121, 128, 153, 157, 158, 173, 179
Orthogonal, 26, 30, 33, 35, 48, 50, 56, 71, 73, 74, 83, 98, 104, 110–112, 114, 116–120, 122, 127, 143, 151, 160, 163–165, 170, 178, 180
    coordinate curves, 118, 127, 144, 145, 159–161
    coordinate system, 35, 49, 73
    trajectory, 46, 151, 162, 175
    transformation, 24, 25, 32, 40
Orthonormal, 49, 71, 74, 104, 110, 123, 143, 145
    basis vectors, 49, 71, 74
Osculating
    circle, 53, 57, 58, 63, 111, 115, 116, 143
    paraboloid, 140
    plane, 50, 54, 57, 61, 63, 116, 150, 154, 155, 160, 164, 167, 169, 170, 175, 177, 179
    sphere, 57–59, 63, 64
Outer product, 185, 186

Parabolic
    cylinder, 15, 19, 39, 114, 140, 173, 175
    point, 94–96, 108, 114, 137–140, 144, 148, 149, 163–165, 179
Parallel, 39, 40, 44, 49, 51, 53, 59–61, 63, 64, 67, 71, 94, 98, 110, 121, 134, 140, 146, 150, 151, 153, 156, 160, 161, 163, 167, 168, 177
    propagation, 59, 60, 64, 161
Parallelepiped, 40
Parallelism, 59, 64
Parallelogram, 83

Parallels, 12, 13, 39, 138, 162, 169
   of latitude, 13
Parameters plane, 25–28, 67, 81
Parametric
   curve, 26
   line, 26
Partial
   derivative, 5, 6, 21, 33, 35–37, 42, 76, 78, 80, 81, 85, 89, 91, 97–99, 105, 108, 109, 125, 127, 132, 133, 159, 181, 182, 185
   differential equation, 92
Patch, 7, 20, 27, 41, 67, 69, 81, 102, 103, 106, 118, 128, 136, 148, 160, 163, 169, 170
Perimeter, 135
Periodic, 45, 46, 60, 168
Permutation tensor, 7, 38, 48, 49, 74, 83, 84, 129, 130, 162, 181, 185, 186
Perpendicular, 13, 48, 50, 66, 75, 90, 94, 114, 116, 138, 163
Plane
   curve, 12, 16, 18, 22, 46, 54, 57, 60–62, 67, 125, 146, 150–152, 162, 167, 168
   surface, 29, 39, 107, 114, 119, 141, 156–158, 172, 179
Polar coordinates, 7, 25, 60, 106
Pole, 33, 69, 70, 134, 157, 185
Polygon, 19, 26, 135, 136, 148
Polygonal
   arc, 20, 27, 106
   decomposition, 19, 26, 40, 136
   plane fragment, 20, 106
Polyhedron, 5, 19, 20, 25, 40
Polynomial equation, 144
Positive
   curvature, 32, 102, 114, 118, 126, 128, 129, 135, 140, 147, 163
   definite, 23, 29, 40, 41, 77, 88, 92, 107, 113, 122, 124, 138, 140, 141, 165
   orthogonal transformation, 24, 25, 40
Principal
   curvature, 7, 55, 118–126, 131, 137, 144, 145, 148, 163, 165, 170, 178, 180
   direction, 55, 94, 96, 108, 119–123, 144–146, 161–163, 165, 166, 169, 170, 172
   normal line, 49, 61, 67, 151, 167
   normal vector, 6, 30, 47–49, 51, 55, 57–59, 61, 97, 110–112, 114, 115, 117, 119, 120, 142, 153, 155, 160, 168
   radius of curvature, 6, 120, 145
Product rule of differentiation, 53, 84, 90, 111, 181
Profile curve, 12, 13
Projection, 13, 26, 69, 70, 112, 116, 154

Pseudo-
   radius, 7, 19, 126
   sphere, 7, 18, 19, 24, 39, 126, 146

Quadratic
   equation, 7, 93, 94, 118, 119, 122, 142, 144, 145, 172
   surface, 13–15, 121, 172, 179
Quadric cone, 172–174

Radius, 13, 22, 31, 39–41, 52, 53, 57, 59, 61, 64, 68, 111, 115, 125, 126, 128, 131–135, 137, 141
   of curvature, 6, 52, 53, 58, 59, 62, 111, 142, 154
   of torsion, 6, 55, 59, 63, 154
Rank
   -0 tensor, 38, 129, 147
   -1 tensor, 76, 172, 182
   -2 tensor, 38, 76, 83, 172, 182
   -4 tensor, 36
   of matrix, 27, 28
   of tensor, 28, 42, 181
Reciprocal systems, 75
Reciprocity relation, 75
Rectangular, 47
   coordinate system, 8, 25, 30, 32, 33, 54, 56, 60, 62, 84, 98, 104, 121, 173, 182, 185
   plane sheet, 69
Rectifying plane, 50, 61
Reference frame, 89, 176
Reflection, 24, 25, 62
Reflective, 176
Regular
   curve, 29, 45, 51, 60, 65–67, 133, 150, 155
   mapping, 45, 68, 69, 81
   point, 28, 30, 41, 45, 47, 66, 71, 88, 97, 104, 133, 136, 150, 160, 164
   surface, 27, 41, 45, 60, 65–67, 71, 102, 173
Relative
   permutation tensor, 7, 38
   torsion, 55
Riccati equation, 57
Ricci curvature
   scalar, 6, 38, 42, 130, 131, 146
   tensor, 6, 38, 42
Riemann
   -Christoffel curvature tensor, 6, 31, 36–38, 42, 52, 69, 85, 86, 107, 124, 129, 132, 146, 172, 177, 184
   sphere, 69
Riemannian
   curvature, 31, 120, 124, 130
   geometry, 31, 41
   space, 31, 33, 37, 59, 65, 76, 105, 155

Right handed system, 48, 49, 51, 73
Rigid motion transformation, 51, 91, 93, 172
Rodrigues curvature formula, 123, 146, 162
Rotation, 13, 24, 25, 51, 172, 173, 177
Ruled surface, 17, 172, 173, 175, 179, 180
Rules
    of differentiation, 181, 183–185
    of transformation, 74, 78, 84
Ruling, 71, 173, 175–177, 179

Scalar, 21, 23, 44, 48, 56, 105, 172, 185, 186
    triple product, 55
Schur theorem, 31
Scroll, 173
Secant line, 45, 50, 58
Second
    curvature, 51
    derivative, 6, 53, 58, 63
    fundamental form, 5, 6, 10, 11, 39, 52, 62, 84, 85, 89–94, 96–99, 106–109, 112–115, 118, 121–124, 128, 131, 132, 138, 140, 143, 144, 147, 148, 163, 166, 170, 176, 180
    fundamental quadratic form, 89
    groundform, 83
    order covariant derivative, 183, 184
    order differential, 90, 107
    order differential equation, 159
Segment, 18, 20, 29, 43, 44, 46, 79, 88, 106, 136, 151, 156, 157, 159
Self-conjugate direction, 165, 166
Semi-circular, 134, 157
Semicolon notation, 5
Similarity transformations, 85
Simple surface, 67, 68, 103, 155, 172
Simply connected, 29, 60, 67, 103, 133, 136
Sine of angle, 83
Singular point, 28, 41, 71
Smooth, 21, 27, 41, 45, 53, 55, 57, 60, 66, 67, 108, 121, 123, 127, 137, 143, 145, 150, 158, 162, 165, 166, 170, 171, 175, 185
Space
    basis vector, 5, 9, 47, 181, 185
    metric tensor, 6, 9, 74, 76–78, 87, 107, 183–185
Span, 50, 66, 103, 111
Sphere, 12, 13, 19, 22, 26, 31, 40, 41, 58, 68, 69, 101, 102, 106, 115, 116, 119–121, 125, 126, 128, 131, 134, 136–138, 141, 143, 145–147, 149, 153–158, 160, 162, 163, 168, 169
    mapping, 101, 102, 109
Spherical
    coordinate system, 105, 106
    curvature, 59, 63
    curve, 154, 168
    image, 101, 102, 109, 127, 163, 168, 170
    indicatrix, 5, 7, 101, 153, 154, 168
    triangle, 134, 148
    umbilical, 102, 141, 146
Square, 20
Stereographic mapping, 69, 70, 103
Straight line, 12, 13, 20, 25, 29, 39, 40, 43, 45, 46, 49, 52, 53, 61, 67, 150, 155–161, 164, 166–169, 171, 173, 176, 177
Stretching, 68, 69, 125
Sufficiently
    close, 158
    differentiable, 21, 24, 81, 146
    smooth, 21, 40, 45, 63, 65, 67, 82, 92, 94, 112, 114, 115, 121–123, 125–127, 132, 143, 145, 149, 159, 162–166, 169, 175, 176, 182
Sum rule of differentiation, 181
Summation convention, 9, 34, 96, 100
Surface
    basis vector, 5, 7, 9, 28, 30, 36, 42, 66, 73–76, 90, 97–99, 104, 105, 107, 108, 118, 130, 144, 181–184, 186
    coordinate system, 74, 75, 77, 84, 104, 105
    coordinates, 6, 9, 10, 22, 25, 27, 29, 30, 43, 45, 66, 67, 71, 74, 76, 78, 80, 81, 85, 87, 89–91, 94, 97–99, 104, 108, 109, 123, 125–129, 131–133, 159, 182, 183, 185, 186
    metric tensor, 5, 9, 38, 42, 73, 75–79, 84, 86, 97, 100, 105, 108, 116, 124, 127, 129, 146, 159, 182–186
    of revolution, 12, 13, 16, 18, 39, 40, 125, 146, 156, 162, 168, 169, 177, 180
Symbolic notation, 47, 56
Symmetric, 12, 34, 37, 38, 42, 62, 69, 76, 77, 80, 83–85, 93, 100, 101, 176, 180, 182, 183
Symmetry, 12, 13, 29, 41, 84, 138, 166, 172, 173, 177, 183
System of
    coordinates, 6, 8, 12, 13, 23–25, 29, 30, 32, 33, 35, 41, 47, 49, 51, 52, 60, 62, 63, 65, 72–78, 84, 87, 88, 98, 104, 105, 110, 121, 125, 127, 159, 160, 172, 173, 181, 185
    differential equations, 57
    linear equations, 122

Tangent
    indicatrix, 153, 168
    line, 45, 49, 50, 58, 61, 63, 71, 113, 151, 152, 167, 175, 177
    plane, 28, 65–67, 69, 71, 72, 93, 94, 102–104, 110, 112, 114, 116, 118, 121, 128, 138, 140, 147–150, 160, 164, 170, 175, 177, 179–181

space, 6, 66, 67, 73, 74, 98, 103, 112, 116, 144, 158, 166
surface, 6, 71, 104, 128, 151, 152, 167, 175, 177–180
unit vector to curve, 47–49, 55, 60, 110, 151, 153, 177
vector to curve, 13, 44, 45, 52, 57, 60, 61, 66, 67, 72, 82, 103, 110, 133, 153, 161, 184
vector to surface, 65–67, 72, 158
Tangential component, 101, 142, 143, 164, 182
Tensor
  analysis, 8, 74, 131
  calculus, 1, 9, 10, 93, 184, 185
  notation, 7, 30, 38, 45, 47, 54, 60, 71, 73–75, 84, 91, 94, 97, 98, 100, 104, 105, 108, 109, 119, 131, 159, 162, 170, 179, 183, 186
Tetrahedron, 25
Theorema Egregium, 101, 132, 133, 136, 148
Third
  curvature, 51, 61
  derivative, 63
  fundamental form, 5, 6, 94, 96, 108
Topological
  image, 43, 46
  invariant, 136
  property, 69, 148
Topologically-equivalent, 137
Topology, 8, 12, 68, 133
Torsion, 7, 48, 49, 51, 52, 54–57, 59, 61–63, 92, 150, 152, 154, 164, 165, 167, 168, 170
Torus, 13, 15, 19, 20, 31, 39, 40, 68, 103, 128, 129, 131, 136–138, 147, 148
Total
  curvature of curve, 51, 145
  curvature of surface, 6, 51, 110, 120, 128, 133, 135–137, 145, 147, 148
  derivative, 33, 182, 185
Trace
  of curve, 43, 60, 65
  of matrix, 6, 65, 85, 131
  of surface, 65, 101, 102
Tractrix, 18, 19, 24, 40
Transformation of coordinates, 6, 12, 23–25, 27–29, 32, 33, 40, 44, 46, 51, 55, 74, 78, 80, 84, 85, 91, 93, 104, 105, 124–126, 131, 138, 160, 172, 183
Transitive, 176
Translation, 24, 25, 51, 172, 173
Transpose, 85, 88, 91
Triad, 49–51, 73, 112, 123
Triangle, 20, 134, 135
  inequality, 29, 41

Trigonometric function, 46
Triply orthogonal system, 162
Type of tensor, 37, 182, 183

Umbilical point, 102, 115, 119, 121, 122, 128, 141, 142, 145, 146, 149, 161, 162, 169, 172
Unit sphere, 41, 69, 70, 101, 153
Unity, 24, 44, 126
  speed, 44
  tensor, 76

Variational principle, 23, 155, 168
Vector notation, 104, 183
Vertex, 114, 136, 141, 160, 172
  of polyhedron, 5, 19, 20, 40, 136

Weingarten equations, 52, 97–100, 108, 109

Printed in Great Britain
by Amazon